SpringerBriefs in Applied Sciences and Technology

SpringerBriefs present concise summaries of cutting-edge research and practical applications across a wide spectrum of fields. Featuring compact volumes of 50 to 125 pages, the series covers a range of content from professional to academic.

Typical publications can be:

- A timely report of state-of-the art methods
- An introduction to or a manual for the application of mathematical or computer techniques
- A bridge between new research results, as published in journal articles
- A snapshot of a hot or emerging topic
- An in-depth case study
- A presentation of core concepts that students must understand in order to make independent contributions

SpringerBriefs are characterized by fast, global electronic dissemination, standard publishing contracts, standardized manuscript preparation and formatting guidelines, and expedited production schedules.

On the one hand, **SpringerBriefs in Applied Sciences and Technology** are devoted to the publication of fundamentals and applications within the different classical engineering disciplines as well as in interdisciplinary fields that recently emerged between these areas. On the other hand, as the boundary separating fundamental research and applied technology is more and more dissolving, this series is particularly open to trans-disciplinary topics between fundamental science and engineering.

Indexed by EI-Compendex, SCOPUS and Springerlink.

More information about this series at http://www.springer.com/series/8884

Christian Lexcellent

Artificial Intelligence versus Human Intelligence

Are Humans Going to Be Hacked?

 Springer

Christian Lexcellent
Mécanique Appliquée
FEMTO-ST
Besançon, France

ISSN 2191-530X   ISSN 2191-5318   (electronic)
SpringerBriefs in Applied Sciences and Technology
ISBN 978-3-030-21443-2   ISBN 978-3-030-21445-6   (eBook)
https://doi.org/10.1007/978-3-030-21445-6

This Springer imprint is published by the registered company Springer Nature Switzerland AG
The registered company address is: Gewerbestrasse 11, 6330 Cham, Switzerland

Contents

Chapter 1
Issue of the Debate

Abstract One introduces the point of view between advocates of Artificial Intelligence and philosophers who are more faithful to Human Intelligence.

The title is derived from a quote from Yuval Noah Harari, "Technologies for hacking people are available" [5].

The so-called artificial intelligence—AI—is currently arousing such a craze that it is difficult to stay away from it.

At the request of the Prime Minister Édouard Philippe, the mathematician and Member of Parliament Cédric Villani and his collaborators produced a 235-page report entitled "Give meaning to artificial intelligence", as if there was room for doubt about that! Laurent Alexandre wrote a book in French, entitled "La guerre des intelligences: Intelligence Artificielle versus Intelligence Humaine" ("The war of Intelligence: Artificial Intelligence versus Human Intelligence") [1]. He wants to boost human intellectual abilities and simply advocate for a revolution in our educational system. Let us say that for him, developing AI is essential if one does not want to die silly (or indeed does not want to die at all!). In short, we will come back to his rather questionable remarks later on. Emmanuel Fournier tackles the problem rather from the viewpoint of cognitive neuroscience, in which the brain is given ever greater space (if we extrapolate to AI), and is henceforth believed to regulate not only our thought but also our emotions, our doubts, our loves, etc. (in this case it is necessary to turn to human intelligence). In his book "Insouciances du cerveau", preceded by "Lettre aux Ecervelés" [4], Emmanuel Fournier tackles the problem with humor.

In the same way, in his famous book "La mémoire, l'histoire, l'oubli" published in 2000 [7], Paul Ricoeur addresses the issue of sensitivity in neurosciences, which has a technical, sometimes even AI approach of the brain. In "Human Memory and Material Memory" [6], Christian Lexcellent addresses the same issue concerning neurosciences, which answer "everything"related to the functioning of the brain and

C. Lexcellent, *Artificial Intelligence versus Human Intelligence*, SpringerBriefs in Applied Sciences and Technology, https://doi.org/10.1007/978-3-030-21445-6_1

in turn of memory. Almost identically, Emmanuel Fournier challenges the power of neurosciences to say everything about self and thought.

In this context, Laurent Alexandre advocates for "the massive use of neurotechnologies at school so that tomorrow our children can compete with robots" [1].

Cédric Villani was entrusted with a mission, namely, "define a strategy so that France moves to the forefront of AI and derives the greatest possible economic benefits from it" [3].

Well, the outlines of the debate are laid between AI advocates like Laurent Alexandre and to a lesser extent Cédric Villani on the one hand, and philosophers like Paul Ricoeur and Emmanuel Fournier, on the other hand, who are more faithful to human intelligence as it is today, if I may say so.

For Henri Atlan [2], "it is a question of sending back to back a neo-materialist ideology for which everything boils down to the brain or to DNA, and its opposite, an irreducible idealism, allergic to any mechanistic interpretation in the living, sometimes for the benefit of creationism or 'design intelligence' (the intelligent design postulates the presence of a divine design coiled up in the molecules)" [8].

However, there exist smart mechanisms or smart machines, e.g., the wheel that gave birth to the bicycle or the washing-machine that brought huge relief to many women around the world.

Today, humanity has become capable of self-destruction and, at the same time, it is making great strides towards "the augmented man". No matter how one imagines it, humans have always sought to improve themselves to overcome their difficulties. Thousands of years ago, they wanted to make equal with their gods and then with one god, the most perfect one of all. Nowadays, they increase their abilities by way of technological objects that are not only outside, but also inside them. Do they want to become "all-powerful" by turning into "living machines"? Will this path not be stopped by the exhaustion of certain raw materials and the degradation of the environment? Is "transhumanism" possible—and is it desirable? See Peter Bu's text: 2018 (http://blogs.mediapart.fr/blog/peter-bu).

In order to enlighten the reader, let us first define the concepts of artificial intelligence (AI) and human intelligence (HI).

Just an anecdote: a cartoon image calls out to us.

"Do we really need robots? The day will come when this question does not even arise and when robots ask what use we are to them".

References

1. L. Alexandre, *La Guerre Des Intelligences: Intelligence Artificielle Versus Intelligence Humaine* (J.C, Lattés, 2018)
2. H. Atlan, *Cours de philosophie biologique et cognitiviste, Spinoza et la biologie actuelle* (Odile Jacob, 2018)
3. S. Fay, D. Nora, Cédric villani: "l'intelligence artificielle sera partout, comme l'électricité". L'OBS **2782**, 27–33 (2018)

4. E. Fournier, *Insouciances du cerveau précédé de Lettre aux écervelés*. éditions de l'éclat (Philosophie imaginaire, 2018)
5. N. Harari, Yuval, *21 lessons pour le XXIe siècle* (Albin Michel, 2018)
6. C. Lexcellent, *Human Memory and Material Memory* (Springer, 2018)
7. P. Ricoeur, *La mémoire, l'histoire, l'oubli* (POINTS, 2000)
8. N. Weill, Le biologiste et philosophe henri atlan relit l'"ethique" à la lumière des récents résultats des sciences cognitives: Le cerveau de spinoza. *Le Monde des livres*, No 22801 (2018)

Chapter 2
Artificial Intelligence

Abstract Artificial intelligence is a set of theories and techniques that develop complex computer programs that are able to simulate certain traits of human intelligence (reasoning, learning, etc.).

2.1 A Definition of Artificial Intelligence

Artificial intelligence: a set of theories and techniques that develop complex computer programs that are able to simulate certain traits of human intelligence (reasoning, learning, etc.).

It can also be defined as "the science of designing machines capable of doing things that require intelligence when they are done by humans" Marvin Lee Minski.

Artificial Intelligence (AI) consists of implementing a number of techniques aimed at enabling machines to imitate a real form of intelligence. AI is implemented in a growing number of fields of application.

The notion was born in the 1950s thanks to the mathematician Alan Turing. In his book "Computing Machinery and Intelligence" [14], he raises the question of bringing machines a form of intelligence. He then describes a test known today as the "Turing Test" in which a subject interacts blindly with another human, and then with a machine programmed to formulate meaningful responses. If the subject is not able to make the difference, then the machine has passed the test and, according to the author, can truly be considered as "intelligent".

With artificial intelligence, humans rub shoulders with one of their most ambitious Promethean dreams, i.e., to make machines with a "spirit" similar to their own. For John MacCarthy (the leading pioneer in artificial intelligence with Marvin Lee Minsky), one of the creators of this concept, "any intellectual activity can be described with sufficient precision to be simulated by computer science, electronics, and cognitive sciences". This is the challenge—even more controversial within the discipline—of these researchers at the crossroads of computer science, electronics,

C. Lexcellent, *Artificial Intelligence versus Human Intelligence*, SpringerBriefs in Applied Sciences and Technology, https://doi.org/10.1007/978-3-030-21445-6_2

and cognitive science. Despite the fundamental debates it provokes, artificial intelligence has produced a number of spectacular achievements, for example, in the fields of pattern recognition, voice recognition, decision support, or robotics.

In the mid-1950s, with the development of computer science, the ambition emerged to create "thinking machines" that functioned similarly to the human mind. Artificial intelligence (AI) therefore aims to reproduce mental activities with the help of machines, in the fields of understanding, perception, or decision. By the same token, AI is distinct from computer science, which processes, sorts, and stores data and their algorithms. In the present case, the term "intelligence" has an adaptive meaning, as in animal psychology. It will often be a question of modeling the resolution of a problem, which can be new, by an organism. While expert system designers want to identify the knowledge needed for professionals to solve complex problems, researchers working on neural networks and robots try to draw inspiration from the nervous system and the animal psyche.

The distinction between strong AI and weak AI comes from an article by John R. Searle (1980), whose French translation was made available in 1987 [12].

2.1.1 A Definition of Weak AI

Artificial Intelligence is called "weak" when it only reproduces a specific behavior, but not its operation. In other words, the machine does not understand what it is doing.

Weak AI essentially aims to sum up the result of a predicted specific behavior as faithfully as possible, using a computer program, without any form of improvisation. It is actually a system that mimics smart behavior in a specific area.

The machine seems to act as if it were smart. It can simulate reasoning, learn, and solve problems.

"Eliza", the dialogue agent that simulates exchanges in psychotherapy, is a typical case of this kind of software. This computer program, written by Joseph Weizenbaum between 1964 and 1966, simulates a psychotherapist. It rewords most of the patient's statements into questions to ask him/her. The fact that a human should speak to a machine without being aware of it has been considered as a criterion in Artificial Intelligence.

2.1.2 A Definition of Strong or Ascending AI

Strong AI cannot only reproduce thinking skills and intelligent interactions (analyze, reason, perform rational actions), but also have awareness, emotions, and understand its own reasoning. Some algorithms already make it possible to understand these notions.

The program demonstrates the impression of real self-awareness, true feelings. The machine is therefore thought to be able to have some hindsight on what it does. As of now, things are only at the stage of ambitions and projects, but already opening the door onto ethical and moral questions.

Today, it seems almost impossible to model consciousness, regardless of the complexity of the system. But research is already under way and is developing at a high speed. The milestones are already laid in "In Principio" (a blog on strong artificial intelligence).

Strong AI could solve complex problems, whatever the environment, with a higher level than human intelligence or an equal one.

Strong AI is associated to robots that show highly advanced autonomy properties. It integrates computer-generated imagery systems called virtual reality whose applications are multiple today.

Artificial Super Intelligence (ASI) is related to the power of machines that is augmented and distributed more easily than in a human brain that has more limited input–output, storage and processing capabilities.

At this stage, computer intelligence exceeds human intelligence in all fields, including creativity and social agility. AI overcomes human intelligence when it reaches access to singularity (a germ capable of self-understanding and self-improvement).

For example, in the film HER (directed by Spire Onze in 2014), the hero falls in love with "Samantha", an intelligent, intuitive and surprisingly funny female voice.

Singularity corresponds to a point when the evolution of Artificial Intelligence will be such that computers will be able to create by themselves more powerful and more effective machines than those constructed by humans.

Many books address singularity, including "The Singularity is Near" by Kurzweill [3]. "Singularity—An analysis" by the Australian philosopher David J. Chalmers, notably proposes to first test ASI in an entirely virtual environment disconnected from the real world so as to measure its capacities.

We must understand ASI as a mixture of all sorts that combines the fields in which man is already overcome with those in which he is not overcome yet but will soon be.

In practice, AI already exceeds human capacities as regards memory. The already outdated expert systems or more recent systems like IBM Watson deal with large volumes of information, demonstrating capabilities inaccessible to any human, no matter how gifted.

However, in the context of trans-humanism, we must stop at the concept of "conceiving one's child". If AI tests show that it is likely to be born mentally or physically disabled, it is necessary to resort to abortion! This is consistent with the choice of the best school curriculum (despite school maps) for the success of one's child, chosen by a certain upper middle class including many teachers who know which routes to follow.

2.1.3   A Brief History of Artificial Intelligence

Please note that a 10-page summary (two pages each Wednesday of July–August 2018) article was published in "Le Monde" by Larousserie [4] (a scientific journalist) on the genesis of artificial neural networks. The story is told as a saga and is quite exciting.

2.1.3.1   From the Perceptron to the First Macintosh, the Prehistory of a Revolution

All started well: "The US military has revealed an embryo of what could be a machine that works, speaks, sees, writes, reproduces itself and is self-aware," the New York Times wrote on July 8, 1958. This article describes the Perceptron, created by American psychologist Frank Rosenblatt (1928–1971) in Cornell Aeronautical Laboratories. This $ 2 million machine at the time was as big as two or three refrigerators, and springs of electric wires sprang from it. The scientist promised that his Perceptron would read and write one year later; it actually took more than 30 years. Frank Rosenblatt was then a psychologist nurtured for more than 10 years with the concepts of cybernetics and AI. He designed his Perceptron based on the works of Warren McCulloch (1898–1969) and Donald Hebb (1904–1985). The former proposed "artificial" neurons inspired from their biological equivalents, and endowed them with mathematical properties. In 1949, the latter provided the rules for these formal neurons to learn, like the brain that learns by trial and error. A calculating unit (a neuron) is active (1) or inactive (0) according to the stimuli that it receives from other units to which it is connected, forming a complex dynamic network.

The authors showed that then their connected system could theoreticallly perform logical operations, "and", "or", and thus perform any calculation.

This innovative way triggered one of the first quarrels in that field. On one side were the "connectionists" and their networks of artificial neurons, and on the other side were the proponents of the "classic" machines, our current PCs.

A Perceptron is also a machine capable of learning, especially to recognize patterns or to classify signals.

The enthusiasm plummeted in 1969 with the publication of the book "Perceptrons" by Seymour Papert and Minsky [10]. These two authors demonstrated that the way in which the Perceptron was built could only solve "simple" problems.

In the meantime, two new currents were born and clashed at Stanford University. Two new "AIs" clashed, according to the terminology used by the journalist John Markoff in his book "Machines of Loving Grace: The Quest for Common Ground Between Humans" [7].

John Mc Carthy (who left the Massachusetts Institute of Technology (MIT) for Stanford University) supported a different version of AI from that of neural networks. The other AI, for "augmented intelligence", was a new approach proposed by Douglas Engelbart who belonged to an independent institution at Stanford University.

According to John Markoff, John Mc Carthy did not like this NLS (oNLine System) considered as too "dictatorial". He had his own concept of AI, which is called symbolical. It was designed to imitate the reasoning of the brain by chaining the rules and symbols logically to achieve at least cognitive tasks.

In 1970, a colleague of Minsky's even asserted in Life magazine that "within three to eight years, we will have a machine with the general intelligence of an average human, I mean, a machine that will be able to read Shakespeare, grease a car, tell a joke, fight."

In 1973, the "Longhill" report (issued in England) threw cold water on the subject: "Most researchers in AI and in the related fields confess a sense of disappointment with what has been achieved in the last twenty-five years (...). Nowhere have the discoveries made so-far produced the promised major impact".

In summary, Douglas Engelbart had won. In January 1984, Apple released its first Macintosh.

The victory of the symbolic vision (Minsky, McCarthy) over the connectionist vision had fizzled out.

But it was preparing its resurrection.

2.1.3.2 Scientific Literature Reawakens Artificial Neural Networks

AI beat humans hollow at the game of GO, takes the wheel of their car, replaces them at work, but could also better heal them.

Thanks to the rising lords of the networks (Yann Le Cun, 58, considered as one of the inventors of deep learning, Françoise Souli-Folgelman, a mathematician, ...) the beginning of the 1980s saw artificial neurons rise up again.

A prospective symposium held in Cerisy (Manche) in 1981 had a quite explicit title for AI: "Self-organization, from physics to politics".

The renowned participants, i.e., Henri Atlan, Cornelius Castoriadis, Edgar Morin, René Girard, Isabelle Stengers, Francisco Valera, Jean-Pierre Dupuis, et al. talked about morphogenesis, cellular automata, emerging systems, complex systems, cybernetics, etc. According to Yann Le Cun, the participants were not familiar with the literature of the 1960s on neural networks. But they spoke of "networks of automata", which was close to these issues.

In 1985, a summer school organized in Les Houches (Haute-Savoie) was devoted to disordered systems and organization in biology.

A physicist named John Hopfield from CALTECH University and Bell Laboratories was present at Les Houches. In 1982, he had published an article relating spin glass (an assemblage of small magnets) to neural networks.

In addition to two future US leaders in the field, Geoffrey Hinton and Larry Jackel, in Japan 82-year-old Kuhido Fukushima developed his own ideas around neural networks on his own, during the "winter" of his field. "Each month, at the beginning of the group's creation in 1965, we invited a neurophysiologist and exchanged between us, combining the skills of engineers, psychologists, and neuroscientists. I was fascinated by the mechanism of information processing by the mammalian brain."

As early as 1979, he programmed a network of artificial neurons inspired by biology, more complex than the Perceptron, with learning abilities, that he called Neocognitron.

Finally, Jurgen Schmidhuber, 55 years old, "dreams of building a smart machine that will learn by itself what will allow it to retire!" Like Yann Le Cun, Jurgen Schmidhuber likes to delve into university libraries because "the inventor of an important method must rightly be credited for it".

2.1.3.3   Fitted with Memory, Artificial Neural Networks Rise Up Again

At the end of the 1980s, the mood was rather good in the laboratories interested in a new way of making smart computer programs. In the United States, France, Switzerland, and Japan, people grew enthusiastic about the feats and prospects of the artificial neural networks invented 40 years before.

Facebook and Bell Labs "picked up the best". "The intellectual vibe at Bell Labs was incredible, the scientific ambitions unlimited, the material resources huge, I had a Sun4 computer for myself, identical to the one at the University of Toronto, where I did my post-doc, which was shared with 30 people!", adds Yann Le Cun who is now a professor at New York University and the director of Facebook's artificial intelligence lab.

Out of passionate exchanges emerged the idea of the NIPS (Neural Information Processing System) conferences that started in 1987 in Denver (Colorado), whose audience rose from a few hundred people to more than 8,000 in December 2017 (out of which nearly 10% were Google employees!).

But just before the first NIPS, IEEE had organized an international congress of neural networks with 1,000 people in San Diego.

The basic system under study evoked the functioning of the human brain.

To go further, several researchers had the idea to use several layers of neurons to increase complexity.

One of the key ideas was the back-propagation of a gradient. The gradient is a way of measuring the gap between a good result and the result provided by the machine. Back-propagation consists in stepping back so as to modify the parameters, recalculate the result and see its deviation from the target value, and so on.

A decisive innovation came 10 years later from Switzerland in 1997. Jurgen Schmidhuber and his student Sepp Hochreiter invented a new type of network. The invention of the two Germans consisted of saying that the state of neurons depended not only on the stimuli they received, but also on the stimuli from the previous stages, which amounted to providing the network with a memory [5].

2.1.3.4   A Chill on Artificial Neural Networks

It is mostly related to so-far insufficient computing power for some applications.

Success is sometimes short-lived. By the end of the 1990s, everything seemed to be smiling upon mathematical or formal neural networks, the new avatar of AI.

In Paris in 1994, at the Higher School of Physics and Industrial Chemistry, Professor Gérard Dreyfus's team, in association with Sagem, had an unmanned SUV circulate in a known environment. At the same time, the Bell Labs neural network read checks thanks to Yann Le Cun, its designer.

In 1998, Françoise Soulié-Fogelman and Patrick Gallinari published a conference report with the explicit title "Industrial Applications of Neural Networks". On the menu, character recognition, visual diagnosis, robotics, various predictions, management of financial portfolios.

Paradoxically, it was at that time that "gray clouds began to gather" above connectionist IA.

In the core of Bell labs, "between 1991 and 1996, four directors followed one another and we were asked to focus less on research," Larry Jackel recalls. The ATT company owner broke the labs into two, Lucent on the one hand and AT and T on the other.

In fact, industrialists were reluctant to seize these innovations.

Articles on the subject were rejected by certain newspapers under the motive "deprived of interest".

"We kept believing in neural networks, but engineers had decided that they were quixotic ideas. The number of people who wanted to improve these systems had shrunk", said Geoffrey Hinton. At the NIPS conferences, the topic was no longer a success.

Another so-called SMS-learning technique (or support-vector machine) was developed. Basically speaking, it was a sorting machine.

Another reason explained the chill. Supervised learning, as practised by artificial neural networks, presupposed, by definition, having considerable and well-labeled data to train systems. But giant companies such as Google, Facebook, Amazon, Microsoft, IBM, in short GAFA, forced the lock.

It was then, in the mid-2000s, that the Institute for Advanced Research in Canada funded a program called "Neural Computation and Adaptive Perception".

Thanks to this fresh Canadian blood, neural networks were ready to take over against SVMs and other modes of statistical learning.

2.1.3.5   After 60 Years of Ups and Downs, Neural Networks Triumph

The old recipe of artificial neural networks, developed in the 1950s and renamed deep learning was trendy again.

In October 2012, like every year, an image recognition contest called ImagNet was organized. The aim was to recognize thousands of images taken from the Flickr site by assigning them the correct label among ca. 1,000 categories (Persian cat, husky, sea bass, bison, accordion, cradle, tractor, etc ...). The error rate was 25% with SMS, but 16% with deep learning (Geoffrey Hinton Team).

Another sector completely upset by these new artificial brains was automatic translation.

DeepMind won at the game of GO in March 2016.

The program of Alex Graves, who then became one of the first DeepMind (Google) employees was the best to recognize Arabic, Farsi, and French writings.

A challenge consisting of roadsign recognition by neural networks was won by Jurgen Schmidhuber's team.

But the maturity of a field can be measured from its successes as much as from its first failures, and deep learning provided a number of them; for example, black people were mistaken for gorillas!

In March 2016, a twitter account powered by a Microsoft Tay robot grew racist by drawing a little too much inspiration from the tweets it encountered.

In 2017, a Uber car killed a pedestrian.

In September 2017, an algorithm promised to guess people's sexual orientation!

Very soon, fears arose of developing autonomous artificial intelligence that humans would lose control of.

For Gérard Dreyfus of the Espci-Paris-tech, "The smart item is the person who designs the program, for the moment." More concretely and as additional proof of the maturity of the subject, ethical initiatives flourished to regulate these developments.

Let us leave the final word to the sociologist Dominique Cardon "We still have not made autonomous, independent machines ... We educate algorithms by our behavior online or in our cars, for example, and consequently we still make one with the machines".

2.1.3.6 Comments

The classic approach derived from Turing's works requires considerable computing power and spends several megawatts in energy while the human brain requires 30 watts! AI is programmed to solve a specific problem. With ALPHAGO, we remain in the rules of the game of GO.

It is the same story for chess.

We can say that AI is monotask and that its much higher computational means as compared to humans allow it to beat the human mind for a given problem.

As far as driving without a driver is concerned, the program does not foresee that a woman may cross the highway with her bicycle.

The most spectacular problem posed by autonomous vehicles is the modern version of the well-known "trolley problem": in case of an emergency, and in the absence of a human driver, how should the vehicle "choose" between its victims? Is it better to kill two adults so as to save two children? A pregnant woman crossing the road off the crosswalk or an old man in his own right? The driver or five pedestrians?

2.2 Cognitive Sciences

From a restrictive point of view, cognitive sciences include—modern epistemology, which focuses on the critical study of the foundations and methods of scientific knowledge, from a philosophical and historical perspectives—cognitive psychology, whose object is the treatment and production of knowledge by the brain, as well as the psychology of development, when it studies the genesis of logical structures in children—logic, which deals with the formalization of reasoning—various branches of biology (theoretical biology, neurobiology, ethology, among others)—the sciences of communication, which include the study of language, the mathematical theory of communication, which quantifies information exchanges, and the sociology of organizations, which studies the social dissemination of information.

2.2.1 The Project and its Development

AI finds its deep historical roots in the construction of automata, the reflection on logic and its consequence, the development of calculating machines.

The precursors As early as ancient times, some automata reached a high level of refinement. Thus, in the first century AD, Heron of Alexandria invented a wine dispenser, a cybernetic process before there was even a name for it, that is to say equipped with regulatory capabilities, and based on the principle of communicating vessels. Scientists quickly appeared to grow obsessed with designing animal-like or human-looking mechanisms. After the often successful trials of Albert the Great and Leonardo da Vinci, it was especially Jacques de Vaucanson who grabbed people's attention, in 1738, with his mechanical duck, whose motor and excretion functions were simulated by means of fine gears. As for the calculator, it was imagined and then achieved by Wilhelm Schickard (Germany) and Blaise Pascal (France). Around the same time, in his Leviathan the Englishman Thomas Hobbes pushed forward the idea that "any ratiocination is calculus", an idea that supported the project of universal logical language dear to René Descartes and Gottfried W. Leibniz. This idea was made a reality two centuries later by George Boole in 1853, when he created an algebraic writing of logic (a definition of canonical sets, for example). People could then hope to step from the concept of an animal–machine to the technology of a machine–man.

The founder One of the forerunner theorists of computer science, the British mathe-matician Alan M. Turing, launched the concept of AI in 1950, when he described the "imitation game" in a famous article [14]. The question he posed was the following: can a man connected by a teleprinter to what he does not know to be a machine in a neighboring room be fooled and manipulated by the machine with an efficiency comparable to that of a human being? For Turing, AI was therefore a sham of human psychology, as complete a sham as possible.

Formatting AI Turing's succession was taken up by Allen Newell, John C. Shaw and Herbert A. Simon, who created the first AI program, "The Logic Theorist", in 1955–1956. It was based on a paradigm of problem-solving, with the very premature ambition to prove logical theorems. In 1958, at the Massachusetts Institute of Technology (MIT), John MacCarthy invented Lisp (for list processing), an interactive programming language: its flexibility made it the choice language for AI (it was completed in 1972 by Prolog, a symbolic programming language that allows users to shunt step-by-step computer programming).

The development of the GPS (General Problem Solver) in 1959 marked the end of the first period of AI. The GPS program was even more ambitious than the Logic Theorist, from which it was derived. It was based on logical strategies of the "analysis of ends and means" type: it defined any problem by an initial state and one or more target end states, with operators ensuring the passage from one to the other. It was a failure because, among other things, the GPS did not consider the question of the way a human being poses a given problem. Therefore, detractors were more virulent, forcing AI supporters to greater rigor.

2.2.2 Criticisms of the Project

Between a "radical" line, which considers the cognitive system as a computer, and the point of view that excludes AI from the field of psychology, a median stand is certainly possible. It is suggested by three major categories of criticism.

A logical objection It is based on the famous theorem that Kurt Godel stated in 1931 [2]. It highlighted the incompleteness of any formal system (any formal system includes items with very precise meanings and definitions, but whose truth or falsity cannot be proved: they are incomplete). It would then be futile to describe the mind by curtailing it to such systems. However, for some of them, nothing indicates that the cognitive system should not be considered as formal, because if in the wake of the Austrian philosopher Ludwig Wittgenstein we consider that a living being is a logical system in the same way as a machine, one can conceive that the mind is "formal", that it has boundaries, like any machine.

An epistemological objection A number of biologists and computer scientists find classical AI prematurely ambitious. For them, we should first succeed in modeling the functioning of simpler levels of integration of life (the behavior of "simple" animals, information gathering by the immune system, or intercellular communications) before tackling the human spirit.

A philosophical objection For Searle, [13] the human cognitive system is fundamentally meaningful. But the machine has no intention; it has no conscience. A computer can manipulate symbols but cannot understand them. Thus, AI is believed to work on the syntax of reasoning processes (combinatorial rules), not on their semantics (interpretation and meaning).

Hilary Putnam considers that the description of the thought made by AI in terms of symbols and representations is deceptive. For him, such an approach presupposes a preestablished meaning, whereas everything would lie in the interpretation of "external reality" by the mind. The history of ideas thus shows that the notion of "matter" does not mean the same thing for ancient Greece philosophers and for modern physicists. In the same way, many biologists consider that the nervous systems of the different animal species reveal distinct worlds in their environments. AI is therefore thought to ignore this phenomenon of "active construction" of multiple realities by the cognitive system.

Finally, among the things that computers cannot do (1972), Dreyfus [1] emphasizes stricto sensu understanding implies a whole common sense. If AI programs failed to adequately address this issue, this would make them counterfeit—but the same author was quite interested in research on neural networks.

Problem solving For the epistemologist Karl Popper, any animal, as a being adapted to its environment, is a problem solver. If problem-solving is probably not the only salient functional mark of the human mind, it remains a must for the modeler. Two approaches are possible to solve a problem: the algorithm approach and the heuristic approach.

Algorithms and heuristics Algorithms are mathematical procedures of resolution. They are a systematic method, which gives reliable results. But a deterministic heaviness marks their boundaries. By using them to solve certain problems, one can indeed be confronted with the phenomenon of "combinatorial explosion". This last case is illustrated by the Indian fable of "The Sage and the Exchequer". To a wise man who had advised him wisely, the king offered to choose a reward. The old man simply asked that a chessboard be brought and that a grain of wheat be placed on the first square, on the second square two grains, and so on, putting on each new square double the quantity of wheat as compared to the previous one. With a quick calculation, we imagine that the King very quickly regretted giving a gift that proved very expensive, maybe impossible to honor.

In contrast, heuristics is an indirect strategic method that is used in everyday life. It results from the choice, among the different approaches to problem-solving, of those that seem most effective. Although its result is not guaranteed, because it does not explore all the possibilities but only the most favorable ones, it nonetheless allows for a considerable gain of time: when complex problems have to be solved, the use of algorithms is impossible.

Unskilled workers (refugees in Uganda, inhabitants of the poorest villages in Kenya, etc.) annotate images for them to be understandable by computers. No training is needed; only a good sense of observation is required. Wages are around ten euros per day (five times more than for odd jobs). However, the day will come when it is possible to do without data annotation, so that unskilled workers will not be needed any more!

The mathematician Cathy O' Neil warns us about the absence of transparency around AI software. Algorithms create their own reality, and the data they use become their cornerstone. Moreover, there is the myth of mathematics: if we do not understand it, algorithms are double Dutch to us.

The exemplary case of chess Of all games, it was chess that spurred the biggest modeling efforts in AI. As early as 1957, computer scientist Bernstein, devised a program to play two games of chess based on the reflections of Claude Shannon, one of the fathers of the Theory of Information. With a heavily advertised release by the year 1959, the GPS program, in which Simon saw the foreshadowing of a future electronic world champion, was beaten by a teenager in 1960. From that time the whole series of "Chess Programs"was developed, and considered to be more promising. Yet these reflected in a more than deficient way the global heuristics of good players: in these automatic games, regular shots are indeed programmed in the form of algorithms. Contrary to the famous words of a champion of the 1930s: "I study only one move: the good one", the computer does not consider its play in the long run; it successively exhausts all the possible states of a mathematical tree. Its major asset is the "brutal force" conferred by its calculating power and speed. Thus Belle, a computer admitted to the ranks of the International Chess Federation in 1975, could already calculate 100,000 moves per second. Nevertheless, the electronic programs of the time were still systematically surpassed by the masters.

"Deep Thought", an IBM supercalculator, was still beaten flat by the world champion Garri Kasparov in October 1989 (at that time the machine only had a capacity of 2 million moves per second). The Deep Thought project had involved a multimillion dollar budget and hyperperforming computers, and benefited from the advice of the great Soviet-American master Maxim Dlugy. The machines were still algorithmic, but made fewer errors and finer calculations. The Deep Thought team tried to break the billion-moves-per-second threshold, for their computer only calculated around five moves in advance, far less than their human competitor: the connoisseurs felt that this number had to be increased up to more than seven moves. In fact, it appeared that it was necessary to design strategist machines that would additionally be capable of learning. In association with the chess Grandmaster Joel Benjamin for the development of the software part, Feng Hsiung Hsu and Murray Campbell, IBM research laboratories, took over the Deep Thought program—renamed "Deep Blue", and then "Deeper Blue"—by designing a system of 256 processors operating in parallel; as each processor could calculate around three million moves per second, Deeper Blue's engineers estimated that it calculated around 200 million moves per second. Finally, on May 11, 1997, Deeper Blue beat Garri Kasparov by three and a half points to two and a half in a six-game match. Although many analysts believed that Kasparov, whose ELO ranking (2820) was a record and who had proved that his title of world champion was undeniable by defending it successfully six times, had particularly badly played, the victory of Deeper Blue arouse excitement among computer scientists. One of the most astonishing strokes was when, in the sixth game, the machine chose to make the speculative sacrifice of a rider (an important piece) in order to obtain a strategic advantage, a blow hitherto normally reserved for humans.

In 2002, the world champion Vladimir Kramnik only managed to draw a draw against the "Deep Fritz" software program, after eight games, two wins for the human, two for the machine and four draws. Once again, neurons had not taken their revenge on chips.

The human brain: – 80 to 100 billion neurons.
– 10,000 to 20,000 dendrites per neuron.
– 180,000 km of connections
About as many glial cells as neurons to assist neurons.
Nerve influxes transmitted and modulated by synapses.
– 500 million neurons in the intestine
– 40,000 in the heart.

2.2.3 Artificial Neural Networks

In an article published in 1943, Warren McCulloch, a biologist, and Walter Pitts, a logician, [8] proposed to simulate the functioning of the nervous system with a network of formal neurons. These "logistic neurons" were in fact interconnected electronic automata with a 0/1 operating threshold. Although this project had no immediate result, it later inspired Johann von Neumann when he created the classic computer architecture.

A first unsuccessful attempt It was not until 1958 that the progress of electronics enabled for the construction of the first neural network, the "Perceptron", a so-called connectionist machine, by Rosenblatt [11]. This neuromimetic machine, whose analog-type functioning sought to approach that of the human brain, was very simple. Its "neurons", partly linked randomly, were divided into three layers: a layer "specialized" in the reception of the stimulus, or peripheral layer, an intermediate layer transmitting the excitation, and a last layer forming the response. In its inventor's mind, the Perceptron was expected to be able to note down any conversation within a short time and render it to a printer. Many teams worked on similar machines in the early 1960s, seeking to use them in pattern recognition: it was a total failure, which caused the work on networks to be abandoned. These seemed to have no future, although researchers like Shannon believed the opposite.

Current networks In fact, the advent of microprocessors—electronic chips—allowed for networks to resurge in a renewed form in the late 1970s, generating a new field of AI in full expansion called neoconnectionism. The new networks were made of simple processors and no longer had parts with specialized functions. The mathematical tools applied to them were essentially derived from modern thermodynamics and the physics of chaos.

The human brain is characterized by a massive parallelism, in other words, it is capable of processing many signals simultaneously. In networks too, many electronic components—neuromimes—work simultaneously, and the binding of one neuromime with others is expressed by a numerical coefficient called synaptic weight. However, this is far from the human central nervous system, which comprises around 10 billion nerve cells and 1 million billion synapses (or connections). Unlike what happens in the brain, when neuromimes send a signal, they always activate their neighbors and do not have the ability to inhibit them. Nevertheless, these machines are endowed with a self-organization capacity, just like living beings: they do not require a posteriori programming. Memory can survive partial destruction of its network; their learning and memorizing abilities are therefore significant. A microcomputer processes information 100,000 times faster than a network, but the network can perform several operations simultaneously.

2.2.4 Some Applications

Pattern recognition is, along with natural language recognition, one of the areas networks excel at. In order to recognize shapes, a classical robot "calculates" them based on algorithms. All the points of the image are digitized, and then a measurement of the relative distances between the points is made by reflectance (the ratio between incident light and reflected light) analysis. Better yet, the absolute deviation of each point from the camera that focused on the image is measured.

These methods, which date back to the late 1960s, are highly cumbersome and prove ineffective when the object captured by the camera is in movement. Although the network is not very efficient for calculations, it recognizes a form on average 10,000 times faster than its conventional counterpart. Moreover, thanks to the variations in the excitation of its "neurons", it is always able to identify a human face, whatever its changes in appearance. This is reminiscent of the characteristics of human associative memory, which coordinates in a complex manner basic features or information into a memorized global structure. Another similarity with the human cognitive system is to be noted: out of one hundred forms learned in a row, the neuronal computer will retain seven. This is approximately the "size" of human short-term memory, which is six items.

Artificial retinas, which were developed in 1990, will gradually make cameras obsolete as the main sensor used in robotics. Like the cones and rods in the eye, these "retinas" with an analog architecture transform light waves into an equal number of electrical signals, but they still ignore color. Some of them can detect moving objects. It will be possible to miniaturize such bioelectronic membranes relatively shortly.

Finally, formal neural networks are also excellent at detecting ultrasound or thermal variations from a distance.

Using a conventional computer, it is possible to simulate text reading by using character recognition software, an optical reader, and a text-to-speech system that will read the text aloud. But certain neural computers are also able to actually teach how to read. Similarly, when a system is coupled to software with a memory containing twenty or so voices sampled from one language, the network forms an effective computer-assisted teaching system that can correct its students' accent!

Artificial intelligence and education The logo language designed by Seymour Papert (Max Planck Institute) allowed AI to substantially contribute to the educational methods intended for young children. Through straightforward programming, logo encourages the child to better structure his/her relationships to notions of space and time, through games. The key idea of logo is based on the following observation made by Jean Piaget: the child better assimilates knowledge when he/she has to teach others, in this case the computer, by programming it.

Although this computer tool contributes to reducing sociocultural inequalities between certain youths, it is doubtful, despite its promoters' wish, that it can help subjects to acquire concepts considered as the prerogative of their elders by several years. Indeed, Piaget's works indeed show that mental structures are built up according to a relatively well-defined chronology and sequence. No matter how excellent a method is, you cannot teach anything at any age.

2.2.5 Perspectives

Taking into account how difficult it is to perfectly model intellectual activity, some AI practitioners have searched for much more modest but thoroughly successful solutions, particularly in certain applications of robotics: AI without a representation of knowledge.

Around 1970, the theoretical concepts of Marvin Lee Minsky and Seymour Papert on the "Society of the Mind", among others, founded a new AI—distributed AI—also called multiagent AI. The proponents of this approach wanted a number of autonomous agents, whether robots or expert systems, to work together, in a coordinated way, so as to solve complex problems.

After designing sets of associated simple expert systems, distributed AI also reshaped the landscape of robotics, generating AI without a representation of knowledge.

The so-called third-generation robots were able, once turned on, to carry out tasks while avoiding obstacles encountered in their way, without any interaction with the human user. They owed this autonomy to sensors as well as a plan generator, whose functioning was based on the principle of GPS. But to date conventional autonomous robots remain incomplete in their design.

This type of robotics actually seems to be stuck in a stalemate: no significant progress has been made since the beginning of the 1980s.

"Artificial life" At the end of the 1980s, the philosopher Daniel C. Dennett proposed a new possible direction for robotics. Rather than drawing inspiration from humans and mammals, he advised to mimic less evolved beings, but mimic them perfectly. Valentino Braitenberg had already followed a similar path at the Max Planck Institute about ten years before, but his machines pertained to imaginary zoology. On the other hand, since 1985, Rodney Brooks (MIT) has been manufacturing insect-like robots; these are the beginnings of what is called artificial life.

This idea became a reality thanks to the progressive reduction in size of electronic components. A silicon chip, therefore, served as the central nervous system in Brooks's artificial insects: for the moment, the smallest of them occupies a volume of 20 cm3. The researcher started from a simple observation: while invertebrates are not very intelligent, they can do a lot of things, and are also extremely resistant. Working on the modeling of simple stimulus-response reflexes, Brooks elegantly avoided the classical problem of representation of knowledge. In the future, he would like to make his robots work in colonies, like ants or bees; his hopes will only come true if engines can be further miniaturized. Ethology, the science of animal behavior, thus made a remarkable entry into the world of AI.

From Google to Microsoft as well as Apple, IBM, or Facebook, all major companies in the computer world are now working on the issues of artificial intelligence by trying to apply it to a few specific fields. Each has set up networks of artificial neurons made up of servers so as to deal with heavy calculations within huge databases.

2.2.6 Artificial Intelligence for the Prediction of Giant Waves

Source: CNRS Hebdo Number 468 of 13/12/2018.

Oceans are animated by very large, unpredictable and destructive waves that instrumentation cannot fully measure. These waves are called "left rogue waves" right. When they rise under a boat, they can lift it and destroy it.

Researchers from the FEMTO-ST Institute and their Finnish colleagues are studying extreme light waves with similar properties in optics. Using artificial intelligence, they identified the spectral characteristics that could be associated to their emergence. These results, published in the journal Nature Communications [9], could be transposed to oceanography.

It can be noted that I also belong to the FEMTO-ST institute, but to the "mechanics" department, not to the "optics" department.

These researchers from the FEMTO-ST Institute (CNRS/ Universit Bourgogne Franche-Comté) and the Tampere University of Technology (Finland) may have taken a major step towards the analysis and predictability of these extreme waves based on studies of optic fibers. Extreme light waves, with properties similar to those of ocean waves, may indeed appear when intense laser impulses are propagated in optic fiber systems.

"Using thousands of simulations, the researchers formed a network of artificial neurons so as to identify the properties of extreme optical waves over time, only based on their spectral content, which is easier to measure. Thus, artificial intelligence accurately identified spectral characteristics invisible to the naked eye to predict the maximum intensity of a freak wave. These works pave the way for a better understanding of the physical phenomena associated to the occurrence of extreme waves, as well as their predictability. Beyond a possible transposition to oceanography, these results also provide new knowledge to accelerate the development and stabilization phases of high-intensity lasers designed for the industry."

2.2.7 Reservoir Computing

Another more physical and significantly less energy-consuming track is to introduce nonlinear dynamic systems (delayed systems, i.e., memory systems).

The implementation of nonconventional computational methods is therefore essential, and optics offers promising opportunities. We demonstrated the treatment of optical information experimentally by using a nonlinear nanoelectronics oscillator subjected to a delayed reaction. We implemented a neuroscience-inspired concept, called Reservoir Computing, with universal calculating capabilities [6].

References

1. R.L. Dreyfus, *What Computers Can't Do ?: The Limits of Artificial Intelligence* (1972)
2. K. Godel, Thesis: Über formal unentscheidbare sätze der principia mathematica und verwandter systeme" (sur l'indécidabilité formelle des principia mathematica et de systèmes èquivalents) (1931)
3. R. Kurzweill, *The Singularity is Near: When Humans Transcend Biology* (Penguin New York, 2005)
4. D. Larousserie, Les cinq saisons de l'intelligence artificielle. Le Monde (2018)
5. C. Lexcellent, *Human Memory and Material Memory* (Springer, 2018)
6. L. Larger, M.C. Soriano, D. Brunner, L. Appeltant, J.M. Gutierrez, L.C. Pesquera, C.R. Mirasso, I. Fischer, Photonic information processing beyond turing: an optoelectronic implementation of reservoir computing. Opt. Express (2012)
7. J. Markoff, *Machines of Loving Grace: The Quest for Common Ground Between Humans* (Harper Collins Libri, 2015)
8. W. McCulloch, W. Pitts, A logical calculus of the ideas immanent in nervous activity. Brain Theory, 229–230 (1943)
9. M. Narhi, S. Salmela, J. Toivonen, C. Billet, J.M. Dudley, Genty G., Machine learning analysis of extreme events in optical fibre modulation instability. Nature Commun **9**, 4923 (2018)
10. S. Papert, M.L. Minsky, *Perceptrons: An Introduction to Computational Geometry* (MIT Press, 1969)
11. F. Rosenblatt, The perceptron: a probabilistic model for information storage and organization in the brain. Psychol Rev **65**(6) (1958)
12. J.R. Searle, É. Duyckaerts, Document: Esprits, cerveaux et programmes. Quaderni **1**, 65–96 (1987)
13. J.E. Searle, *Les Actes de langage, 1972, éd. Hermann, traduction de Speech Acts: An Essay in the Philosophy of Language (1969)* (Hermann, 1972)
14. A. Turing, *Computing Machinery and Intelligence* (Mind, Oxford University Press, 1950), p. 59

Chapter 3
Artificial Intelligence According to Laurent Alexandre

Abstract An optimistic and pessimistic reading of Laurent Alexandre's book.

3.1 A Rather Optimistic Reading of Laurent Alexandre's Words

The Brain Fight "A world in which intelligence is free of charge and tireless is a major break for any entrepreneur."

These are the introductory words of Laurent Alexandre, a surgeon, a neurobiologist and a novelist.

As far as intelligence is concerned, he distinguishes between the organic brains—human brains—and the silicon brains (silicon makes up electronic chips) of artificial intelligence.

Silicon brains are classified into weak artificial intelligence (AI), distinct from "strong" AI that will one day be aware of itself.

To prepare for the upcoming societal upheaval—at first, the replacement of silicon brains by weak AI, then, in a second step, the overcoming of human brains by strong AI-, Laurent Alexandre advises leaders to get the battlefield ready. "(Human) intelligence is rare, expensive, capricious, and works 35 h a week".

Tomorrow, it will be almost free of charge, ubiquitous, will work 168 h a week and will circulate instantly from Paris to New York or Beijing, he warns.

A growing number of tasks is entrusted to weak AI—and despite this, "they are already better than the best radiologists," explains Laurent Alexandre, convinced that it is not necessary to wait for a Terminator era to start worrying.

How can we fight? How can we make the human brain competitive with silicon brains?

For Laurent Alexandre, the brains that will play their cards right will be complementary to artificial intelligence.

But they will be scarcer and will be worth more.

"A true specialist is worth more than one million dollars a year. Only GAFA can offer this sum today"

Therefore, few actors will actually produce AI.

C. Lexcellent, *Artificial Intelligence versus Human Intelligence*, SpringerBriefs in Applied Sciences and Technology, https://doi.org/10.1007/978-3-030-21445-6_3

In fact, they will buy AI from the giants [therefore GAFA, Editor's note], he says. Google and IBM are already starting to provide this raw material "from the tap".

For business leaders, according to Laurent Alexandre, the aim is to anticipate the capacity of other players to enter their business niche thanks to new technologies.

"The boss of Airbus believes that Google is probably its main competitor to date," the speaker said.

Tomorrow, artificial intelligence will be inserted in autonomous vehicles, but will also monitor the automation of production lines ...

All types of implementation tools can be imagined.

Another challenge is to manage company organization between organic brains and silicon brains.

"What will happen in companies is that AI and HRD will merge. We will manage silicon and artificial intelligence together and synergistically," Laurent Alexandre predicts.

To remain in the fight, businesses must be evaluated on a daily basis. "Managers of SME-type businesses are not prospectivists. But during the phases of major technological change, a reflection on the future will turn out to be determining," the novelist insists.

He suggests establishing a reflection plan, by 2020–2025, to plan what will come next.

One can also read: Technology at the service of the firm of the future. At the dawn of a revolution. "Artificial intelligence will replace millions of jobs, but the whole lot of studies on this question is touchingly naive," the surgeon says.

Because this is without taking emerging jobs into account. "We do not know what tomorrow's jobs will be like".

"We did not know there would be YouTube, BlaBlaCar ...", he warns.

For him, this transition is to be paralleled with the revolution that fueled the production of the first gasoline-powered cars in the early nineteenth century. The drastic reduction of the number of horses in large cities also fueled the fantasies of mass job destruction. And jobs were indeed destroyed. Water carriers, farriers ... "If all the naive ones and the pessimists had carried out their studies in 1895, they would also have added these threatened jobs," says Laurent Alexandre. But other jobs emerged: "The automobile sector created nearly 10 million new jobs, including tourism, road exploitation and car exports," he says.

While expecting this uncertain future, is there a solution to save humanity and prevent it from being disqualified by advances in technology and artificial intelligence?

For the neurobiologist, everything is in the training of young brains. "We need to modernize the school, appoint more competent people, maybe even the best, and overpay them. Teaching should be the highest-paid job in the public service, with high-performance teachers," he explains.

Microprocessors in the head?

He cites Elon Musk, the American billionaire entrepreneur responsible for the success of PayPal, Tesla, and SpaceX, among others. "Elon Musk, with the launch of Neuralink, wants to put microprocessors in the heads of our children."

We are heading for a neuro-enhancement society. "If the kids from California or Singapore do it, we'll have to do it too," he says, adding, "Our Earth will disappear in 2 billion years because the Sun is starting to heat up. Oceans will turn into water vapor in not so long.

"Democracy cannot survive IQ inequalities in a society led by strong artificial intelligence."

"If we do not find a definitive way for us to be over-qualified as compared to strong AI, we will disappear," he concludes.

3.2 A Pessimistic Reading of Laurent Alexandre's Remarks

"Dr. Laurent Alexandre announces the apocalypse that is lurking ahead of us, i.e. congenital stupidity. The IQ of French people, he says, dropped by four points in twenty years' time. It is huge, it is frightening, and several western countries are experiencing the same collapse, while—how dreadful!—the IQ of Asian people is rising at full speed. Please note that the presumed 4-point IQ drop comes from a study conducted on only 79 people! His explanation? "Women with a university education have fewer children than uneducated ones" [5].

According to his words, "The more talented an Englishwoman is, the fewer children she has; the less well endowed with cognitive abilities she is, the more she breeds" [1]. He proposes a solution "The Social Security should fully refund oocyte cryopreservation for female scientists so as to enable them to have babies late, after their Ph.D.".

"The same kind of fantasies had roamed the minds in the nineteenth century. In the United States, the American elites were alarmed by the progress of medicine, which would allow morons or alcoholics to breed like rabbits. The story of the Jude Family, a family issued from a simple-minded couple that had generated several dozen half wits over seven generations, was told alarmingly ... Transhumanists only reactivate this fear. They see themselves as today's and tomorrow's supermen. They want to strengthen their position against us, the 'chimpanzees of the future', who are already in such great numbers" [5].

Laurent Alexandre recalls that "intelligence is the means humanity has been endowed with by Darwinian evolution to survive in a wild environment". He advocates for "the revolution of the transmission of intelligence. In the next few years, the school will certainly go through accelerated modernization under the effect of digital technologies, but it will actually be the last glimmers of an institution doomed to end up on the historical shelf of the curiosities of the past based on rough science."

"From 2035, education will become a 'branch of medicine' using the huge resources of neurosciences to first customize transmission and then optimize intelligence bio-electronically".

According to the author [1], the artificial intelligence machine is already born! "The brain is indeed a computer "made of flesh", but a very complex one, of a different nature than integrated circuits. Thanks to the interconnection of its billions of neurons (around 93 billion), it is able to apprehend unknown situations, to invent, to react to the unexpected and to "reprogram itself" permanently". With deep learning, a first step has been taken.

There are two forms of AI: weak AI and strong AI.

Weak AI is powerful but remains under human control.

Strong AI could develop its own project, and thus run out of the control of its creators.

"To give up AI would be to give up one's smartphone, block the Internet, weaken research, cripple entire sectors of the economy ..." [1] p. 47.

"Man is quickly perceived as the weak link in the face of AI. As Yann Le Cun quite rightly says, "We will soon realize that human intelligence is limited." "The complexity of AI precludes its being evaluated by a human brain: only AI can surveil and evaluate AI."

"Neuroscience specialists will have to be brought into the schools, since the teacher of the future will fundamentally be a 'neurocultor', i.e. a brain-grower" [1] p.54. Let us leave aside educational techniques, teacher—student relationships, "living together", learning, in short all that educated us."

In reference to Bourdieu and Passeron in their now classic book The Heirs [2], as regards the reduction of inequalities, the author assigns a nought to the school system.

"The digital giants bring out cheaper industrial brains than those developed 'through crafsmanship' by National Education".

"The most complex machine in the universe weighs 1,400 grams: it is the brain."

"The school never really succeeded in developing intelligence as it had imagined."

"The demonstration that thought needs nothing more than a network of neurons crossed by nerve impulses is recent."

"The universal income must be forbidden as forcefully as incest" [1] p.143.

"The universal income could quickly 'shrink' our brains by drastically reducing brain plasticity."

"Do we still need classrooms?"

"The era of ideology and of the healers of educational techniques will end, to make room for the era of statistical proof" [1] p.160.

"From around 2030, education will come out of the DIY age to become an exact science."

"Learning becomes a true science based on the objective observation of the brain and its modes of response [3].

In the same vein, here are the words of Ray Kurzweil, chief engineer of Google: "From the 2030s, thanks to the hybridization of our brains with electronic nanocomponents, we will have godlike power." "In thirty years, humans will be able to fully upload their minds to computers to become numerically immortal."

"The desire to be neuro-improved may seem widely shared, due to social conpliance, but its realization is only possible for a few. The risk is therefore great to lead to an 'improved' social class consisting of a small minority of well-informed individuals with sufficient financial resources to access them".

This raises the following question: who will be equipped with these chips and who will not, knowing that the hybridization of our brains is very expensive?

It is also the project of manufacturing superhumans who will dominate the world and who will be more Chinese than European.

"The hope of repelling the agonizing prospect of death is so powerful that the greed for the technologies able to help us is insatiable." [1] p. 243.

"Singularity is close," Kurzweil wrote in 2005 [4]. Singularity is that moment when machine intelligence will overpower human intelligence.".

"AI will drive us to accept the transhumanist vision. This will notably lead to a general IQ increase, and then to the possibility of partly becoming machines. [1] p. 267.

A motto "After Black-White-Arab-, Neuron and Silicon".

"My opinion is that we must find a way to get to the front of the stage (to be part of the 100 Frenchmen or Frenchwomen who count in AI), but at what cost?

At the cost of a vision on the fringe of delirium, of a decomposed society with an elite super-boosted by neural implants, surrounded by a horde of congenital half wits".

References

1. L. Alexandre, *La Guerre Des Intelligences: Intelligence Artificielle Versus Intelligence Humaine* (J.C, Lattés, 2018)
2. P. Bourdieu, J.-C. Passeron, *Les héritiers* (Editions de Minuit, 1964)
3. M. Burns, *The Neuroscience of Learning: Brain Fitness for all Ages* (Scientific Learning, 2011)
4. R. Kurzweill, *The Singularity is Near: When Humans Transcend Biology* (Penguin New York, 2005)
5. J.-L. Porquet, Plouf: submergés par les crétins. *Le Canard enchaîné*, Mercredi 7 mars: 5 (2018)

Chapter 4
The Villani Artificial Intelligence Report, January 2018

Abstract A Cedric Villani report about Artificial Intelligence performed in January 2018.

4.1 Introduction

Within the framework of the Artificial Intelligence Summit "AI For Humanity", Emmanuel Macron presented the government strategy for artificial intelligence at the College de France on Thursday, March 29, 2018. Here is what is to be remembered. In the introduction to his very long speech, the President of the Republic laid the four cornerstones of this strategy:

– "Create a context that will consolidate the ecosystem of artificial intelligence in France and in Europe, and in particular a true network of research and experimentation as regards talents"

- "Start a determined open data policy to bring out the emergence of artificial intelligence champions or encourage their development in France"
- "Have both a financing and a project strategy, French and European public strategies, so as to allow us to develop and accelerate our presence and fully succeed in international competition, from health to mobility"
- "Think the terms of a political and ethical debate that artificial intelligence is feeding worldwide, whose terms we must lay because we need to articulate reflection, rules, and a common understanding".

4.2 The 10 Key Messages of the Report

– Promote the emergence of a European data ecosystem.

To provide relevant results, AI must be based on large databases, from which programs can "learn" and draw correlations. This is why in this report Mr. Villani

© The Author(s), under exclusive license to Springer Nature Switzerland AG, part of Springer Nature 2019
C. Lexcellent, *Artificial Intelligence versus Human Intelligence*, SpringerBriefs in Applied Sciences and Technology, https://doi.org/10.1007/978-3-030-21445-6_4

advocates for "an offensive datum policy aimed at promoting access to data, sharing it and circulating it", particularly at the European level. In practical terms, the aim is to "reinforce" the policy aimed at opening up public data, but also private sector data. "In some cases, the public authorities could impose opening up certain data of general interest," the document says. The MP notably refers to the health sector, for which he would like to create "a platform for accessing to and sharing relevant data for research and innovation".

Create a research network of excellence in AI Develop research. Another priority of the report is to develop AI research: the number of students in the sector is highly insufficient, and young experts trained in France are often recruited by the large Web companies, which have considerable resources to attract them. The goal is to triple the number of persons trained in AI by 2020. To make French research more attractive, the report proposes to increase the number of scholarships, but also to double salaries at the beginning of careers. Mr. Villani calls for the creation of "an AI research network of excellence", which would be composed of interdisciplinary institutes spread across the whole territory, bringing together French and foreign researchers. They would also aim to bring research closer to the corporate world in order to promote technology transfer. These institutes are imagined as "free AI zones", with "a drastic relief of daily administrative procedures, substantial wage supplements, aids for the improvement of the quality of life". Finally, for Mr. Villani, this investment in research should also help equip researchers. He proposes to put a supercomputer and a private cloud allocated to AI at their disposal.

– **Concentrate the economic and industrial effort on 4 priority areas: health, transport, ecology, and defense**
– **Structure the support to innovation in big challenges and carry out trials**
– **Create a Public Lab of Work Transformation**
– **Try a social dialogue at the value chain level to finance professional training**
– **Triple the number of people trained in AI by 2020**
– **Provide for the means to transform public services thanks to AI**
– **Integrate ethical considerations at all levels, from AI design down to its impact on society**
– **Support a bold policy aimed at increasing the number of women in the AI sector**
=> **1.5 billion euros allocated to the budget/13 billion by Google/++ in China.**

4.3   Health

Radiology Detection of very fine visual anomalies in MRI or CT scan to assist doctors in their work.

Skin cancer Identification of melanomas by 3D imaging.
   Connected objects, preventive medicine, mass sequencing of the genome, Automated medical analyses.

Watson used in medical diagnosis

C. Villani:

"The fact that diseases are better detected and treatments cheaper is great. The fact that insurance companies begin to charge incremental rates based on confidential information, that those who suffer the most serious diseases must pay more and more, is obviously not what we want".

4.4 AP-HP Hospital Tests Artificial Intelligence Capable of Replacing a Doctor on the Sly

The Quotidien du Médecin reported on 10th April 2018 that a medical expert system had been tried on patients for several months.

In decision support in oncology, medical imaging, or dermatology, expert systems have proved better at detecting benign skin patches than conventional techniques.

But this was the first time that AI had been used at this scale, during consultations in general medicine.

It was an optimized version of Watson software, developed by IBM, that replaced the physician during the consultations.

The project, co-piloted by AP-HP and a CNRS unit from Pierre-et-Marie-Curie University, was widely used for general medical consultation.

In order to refine their reasoning, Watson's algorithms are powered by the patient's medical data collected from hospitals and the social security by AP-HP . This can pose a problem of confidentiality of this data.

But a human being is never far away. A doctor supervises the consultations, and validates the diagnosis and the prescription.

At this stage, only patients with mild conditions are sent to these consultations.

Another safeguard is that prescriptions are limited to homeopathy treatments.

In order to reassure the patient, he/she leaves with a medical treatment prescribed by a doctor.

In order to train Watson, researchers made it work on the books by the ex-doctor Martin Winckler, and in particular on his latest work, "Brutes en blanc".

Martin Winkler, then a country doctor, made himself known by his humanity towards the sick, and his experience reported in "The Sickness of Sachs" was a great success.

The advantages of AI lie in the fact that the doctor can concentrate more on administrative tasks because he/she is relieved from certain medical obligations.

The time needed for decision-making can be significantly reduced.

Of course, Watson software does not sleep with its patients!

However, can the effectiveness of a software program replace the relationship between the doctor and the patient, who often needs a true word or simply to be heard?

4.5   What Cédric Villani Proposes About Artificial Intelligence in the Healthcare Sector

The report on artificial intelligence (AI) handed to the government by the Villani mission includes several recommendations for supporting the development of AI technologies in the healthcare sector, identified as one of four priority fields where France must focus its economic and industrial efforts.

The President of the Republic is to announce the strategy favored by the government in this field.

1.5 billion euros in public funds will be spent on AI over the entire five-year period.

"Artificial intelligence will not replace doctors and researchers," Emmanuel Macron said during his speech at the College de France.

Around 15 proposals directly concern the deployment of artificial intelligence in medicine and health.

Medical pre-diagnosis, therapeutic strategy, personalization of treatments, predictive follow-up of the development of a pathology, help with choices in the healthcare pathway, management of hospital fluxes, clinical and fundamental research.

The use of AI in the healthcare sector is protean.

Open up Social Security data. The 2010 winner of the Fields Medal proposed to launch a new specific AI site to go along with the shared medical record (DMP, dossier médical partagé).

The DMP would then be "expanded as a secure space where individuals could store their data, add other data themselves, allow sharing with other actors (doctors, researchers, members of the entourage, etc.) and recover them to create other uses."

To develop the potential of AI in the healthcare sector, France must set up a platform devoted to data access and sharing specific to health research and innovation (initially bringing together medico-administrative data, then genomic, clinical, hospital data).

This tool will be intended to eventually replace the national health data system, a nugget that includes 20 billion lines of service articulated with the database of the social security's program for the medicalization of information systems (PMSI).

This "unique system in the world" must open up so as to be better exploited for innovation purposes, the mathematician argues.

In the same vein, Cédric Villani proposes to "develop an understandable offer of access to hospital databases". "Hospitals inherit of, possess, or are building up molecular databases and clinical annotations," he says.

It would be beneficial to encourage hospitals to organize data science bowls or challenges around datasets.

Reform medical studies.

The scientist also addresses the access routes to medical studies, which he wants to "transform" in order to diversify profits and integrate more students specialized in computer science and AI (creation of a double cursus, recognition of equivalent degrees).

According to him, it is also a question of "putting an end to the logic of competition throughout the university curriculum, which is counterproductive, in order to develop a transdisciplinary coordination and structure the doctor—patient relationships of authority".

Unsurprisingly, he advocates for a better training of doctors and other health professionals in the use of AI, the Internet, and big data.

"This transformation of initial training could take place in the current reform of the first and second cycles of medicine studies," he recommends.

One last, important point: AI requires to "clarify the medical responsibility of health professionals".

"In the absence of the recognition of an independent legal personality for the algorithm and the robot, it would be conceivable to hold the doctor responsible for the use of programs, algorithms and artificial intelligence systems, except in case of construction defects of the machine".

4.6   Transport

Autonomous vehicles (cars, taxis, trucks, buses, ships, etc.), Deliveries, traffic, parking

Ecology: waste, pollution, energy, etc.

The state could play a "pivotal role" in "starting the movement and the structuring of the ecosystem" and becoming "the first customer". This could be enacted through innovation support mechanisms, grants, public orders, but also the organization of "big challenges".

Moreover, the MP believes that France should "take the leadership" as regards the links between AI and the environment: "AI can contribute to reduce all our consumptions and amplify all our actions in favor of the respect and restoration of ecosystems," says the report.

Defense, Intelligence, Cyber Security, Justice, and Predictive Police?
Banks, translation, social networks, etc.

4.7   Beyond Fantasies, What Are the Concrete Problems that Artificial Intelligence Poses?

The recent advances in these technologies pose, as of now, less spectacular but much more concrete questions:

– **Concern for the job market** Experts decipher the benefits of AI and the changes it will produce. "I am convinced that an industrial revolution will be brought about by artificial intelligence that will upset the job market, what we do not know is the speed at which this will occur. What is going to happen today is that people in their

30s, 40s, 50s will lose their jobs and will not be ready for the newly created jobs," says Yoshua Bengio, a researcher specialized in AI. The threat is hovering.

– Programs as racist and sexist as humans A number of AI technologies "learn" from huge databases created by humans, which they must draw inspiration from to make conclusions. Yet, these datasets are often biased. As a result, several programs have already shown that they reproduced human racism or sexism. Thus, when an AI program became the jury of a beauty pageant in 2016, it eliminated most black candidates. Another technology, supposed to make links between words, reproduced certain stereotypes, for example, by associating women to daily chores and men to scientific professions.

AI, therefore, learns from our own prejudices to better reproduce them. How can we improve the situation? If databases are biased, it is also, according to Emmanuel Mogenet, director of Google Research Europe, because they are often incomplete, and not sufficiently representative of minorities. "We must find the spots where we have not collected enough data," he explained to Le Monde in April 2018. "It is a problem that we are currently tackling, which is a huge concern to us, because we want models that do not ignore minorities. We are making progress."

– Hacking the human spirit—and democracy Eric Horvitz said it bluntly: he fears "AI attacks on the human mind". At the SXSW festival held in March 2018 in Austin, Texas, this leading artificial intelligence specialist, director of Microsoft Research Labs, listed the potential dangers of AI that he believes should be addressed right now. What he means by "attacks" has nothing to do with a cyberpunk frenzy about implementing technology in the brain. The danger is much more tangible—and for him it is already here. For example, Eric Horvitz mentions AI programs that can write a tweet "specially designed" for a person. "What does the person tweet about? When does he/she answer? What events did he/she participate in? These data can be used to design a tweet that makes it almost impossible for [the person] not to click. A new step for targeted advertising, but not just that. "Companies use this data to personalize messages, but also to influence people's vote, like Cambridge Analytica does." Eric Horvitz also mentions the risk of "fake news", forged news assembled from scratch, that could benefit from these technologies: for example, programs are currently able to make Barack Obama or Vladimir Putin say what we like on video. An issue that is not specific to artificial intelligence, but these technologies make it possible to automate and simplify these means of influence.

– The spectre of autonomous weapons: participate in research or not? The spectre of autonomous weapons. For the moment, states affirm that the robots used in armies are still remotely controlled by humans (Quentin Hugon/Le Monde). Given the level of AI and robotics technologies, there is nothing technically opposed to the creation of lethal autonomous weapons. Today, armies claim that the machines they use are still remotely controlled by humans, like the drones of the US Army, and that none of them ever makes the decision to fire. But no international regulation currently forbids the use of autonomous lethal weapons, which is the subject of discussions at the UNO. In 2015, more than one thousand people, including many AI researchers, but also personalities like Elon Musk or astrophysicist Stephen Hawking, signed a call for a ban on these weapons. "Artificial intelligence has reached a point when the deployment of such systems will be—materially, if not legally—feasible within a few years, not decades, and stakes are high: autonomous weapons have been described as the third revolution of the war techniques, after gunpowder and nuclear weapons", one could read in this call.

NB: Seeking to create an ideal soldier, obedient to all orders from a distance, the Canadian army (which surely was not the only one) already tried implanting electrodes in the brains of monkeys in the 1960s. This has nothing to do with "cyberpunk frenzy", but with military frenzy to which autonomous weapons respond in another way.

– A new step in surveillance Computer vision has made significant progress in recent years thanks to advances in deep learning. Programs are now able to recognize faces, to distinguish a cat from a dog, and to describe pictures. These innovations more and more apply to video, especially video surveillance. For example, shortly after the November 2015 attacks in the Paris region, SNIF announced that it was trying technologies to detect suspicious behavior from surveillance cameras, based on criteria such as "the change in body temperature, a louder voice or the jerky nature of gestures, which can show anxiety". Coupled with facial recognition technologies, this type of system could, for example, make it possible to detect in real time a person on file abandoning a suspicious package; but also a human rights activist in a dictatorship or a homosexual in a country where this is considered illegal. These systems are still far from working perfectly, and the risk of "false positives" remains high.

– Opaque systems Thanks to artificial intelligence technologies, it is possible to create programs to select CVs, offer medical diagnosis or approve a loan. But a lot of the decisions made by these programs—cannot be explained. In fact, engineers do not know how to trace the multitude of calculations made by the machine to reach its conclusion. Clearly, this means that if your loan application or your CV is refused, no explanation can be provided. An embarrassing finding, which explains among other things why today AI technologies are generally only used to suggest solutions later validated by humans. Explaining the functioning of these technologies based on artificial neural networks is one of the major challenges of AI researchers, who are working on the issue. "Explaining the behavior is very important, it is what determines the acceptability of these systems by society," David Sadek, director of research at Mines Telecom, explained to the Senate on January 19, 2018. In recent months, the controversy about the APB algorithm that arbitrates the higher education choices of baccalaureate graduates—which is not an AI program, but whose code has long remained completely secret—already showed that the opacity of automated systems posed significant problems.

– Important legal issues to be solved "If robots develop, who will be responsible? The question will arise of a compensation in case of damage", asked Jean-Yves Le Déaut, then an MP, at a Senate hearing on January 19, 2018. The question is a concern, even if the law does not seem to be about to change, whether in France or elsewhere. "Automated systems are increasingly going to have to make decisions about situations that engineers have not been able to predict," explained Derek Jink, a law professor at the University of Texas Law School, during the SXSW festival in March. "Who, for example, will be responsible for the acts of autonomous cars? The question often comes up, and haunts insurance agents: if an autonomous car kills someone in an accident, who will be held responsible: the manufacturer, the engineer who developed the AI, the car owner, or the person sitting in the driver's seat? These are urgent questions, while experimental autonomous cars have already traveled millions of kilometers on real roads in the United States.

As for Terminator, you can try again! "Singularity annoys me". In April, Jean Ponce, a researcher in artificial vision at the Ecole Normale Supérieure (ENS), criticized the proponents of this concept which designates the hypothetical moment when artificial intelligence will exceed human intelligence. "I personally do not see any clue indicating that the smart machine is closer to us today than before," he explained at a conference organized by Mogole in Paris. In collective imagination, artificial intelligence tirelessly evokes images of Terminator films, in which intelligent machines declare war on man. But in reality, the vast majority of AI researchers say that they have no idea how a machine as intelligent as humans could be created, capable of natural dialogue, of common sense, of humor, able to understand its environment, and even less so in the form of a humanoid robot. The idea of an AI that escapes the control of its creator also arises wry smiles in the community, who find it hard to understand why some people fear that a program designed to play at the game of go would suddenly want to attack the human race. "It's amazing how excited people are," said Eric Horvitz, director of Microsoft Research Labs, at the SXSW festival. Journalists tend to disseminate an extreme point of view, whereas reality is much more nuanced than that. For him, these "remain very interesting questions, (...) on which we must keep an eye, and we must not make fun by saying that people are crazy". But, he says,"these are very long-term issues, and we need to think about the issues that concern us directly now." Blade Runner: the (fake) lawsuit: if an android is indistinguishable from a human being, does it have the same rights as a human being? "Le Monde" organized a debate in the context of the Monde Festival on Sunday, September 24, 2018 at the Opéra Bastille, in the form of a fictional trial, after Philip K. Dick's short story and Ridley Scott's film.

4.7.1   The 100 French Weighty AI People

Gurus/Pioneers/Accelerators/Integrators/Start-Uppers/Mediators.

How to choose one hundred talents? The multiplicity of French artificial intelligence (AI) talents, from the grizzled pioneers to the ambitious deep learning youngsters could have resulted in the "AI 1,000". More modestly, we retained one hundred people, after a first selection of 200 names collected by the editorial staff of "L'Usine Nouvelle", based on their own knowledge of AI actors and suggestions from many sources: individuals, research institutions, companies, associations,etc. Far from being a ranking, our list aims to provide a representative sample of the French AI ecosystem in order to sketch a portrait of it. Hence the broad spectrum swept, from researchers to financiers, including entrepreneurs, institutional personalities, corporate users, decision makers,etc. The famous names are of course present, but we also wanted to highlight the diversity and renewal of the ecosystem by naming those who often remain behind the scenes and are working to develop AI tools and integrate them into products and solutions.

101, 102, 103 ...

One hundred additional French talents who also make artificial intelligence progress in France and around the world.

Serge Abiteboul, Inria/Florence d'Alché-Buc, Mining Telecom Institute/Moojan Asghari, Women in AI/Emmanuel Bacry, CNRS/Michaël Benesty, Lefebvre Sarrut/Mehdi Benhabri editions, Opecst/Alain Bensoussan, Bensoussan/Charly Berthet cabinet, CNNum ....

Any selection has a part of arbitrariness. Many other artificial intelligence professionals would also have deserved to be mentioned. Therefore this supplementary list does not exhaust the subject, but it testifies to the great wealth of French AI.

4.7.2 Artificial Intelligence and Work

Artificial intelligence will lead to profound transformations of work. To prepare for this, the report screens three sectors—transport, banking, and health—develops evolution scenarios, and proposes courses of action.

To shed light on these debates, Muriel Pénicaud, the Minister of Labor, and Mounir Mahjoubi, the Secretary of State for Digital Technologies, have entrusted France Stratégie with a mission on the impacts of artificial intelligence (AI) on work. This mission is complementary to that entrusted by the Prime Minister to the MP Cédric Villani, who addressed within a wider scope the issues of research, industrial policies, or ethics. The objective is the same: not only provide educational information to avoid fantasies, but also evaluate the forthcoming transformations while identifying the adapted public policies. On the basis of its analysis, the report identifies the following three axes aimed at addressing the issues raised by artificial intelligence at work:

– conduct prospective work on the potential of artificial intelligence at the branch or the sector scale, so as to ensure a good level of information and anticipation of the actors;

– train workers about tomorrow's challenges: train highly qualified workers to produce AI, aware of the technical, legal, economic, or ethical issues that go along with the use of artificial intelligence tools;

– reinforce career path securing schemes for the few sectors or subsectors expected to be strongly impacted by the risk of automation.

Finally, the impacts on working conditions—loss of autonomy, intensification of work, etc.—related to the conditions of deployment of AI tools in work units must not be underestimated.

4.7.3   Academic Researchers and Industrialists Unite to Create the PRAIRIE Institute

A place of excellence is dedicated to artificial intelligence in Paris:

"PaRis Artificial Intelligence Research Institute".

The CNRS, Inria, PSL University, and Amazon, Criteo, Facebook, Faurecia, Google, Microsoft, NAVER LABS, Nokia Bell Labs, PSA Group, SUEZ, Valeo, merged academic and industrial interests and united to create the Institut PRAIRIE in Paris, whose objective is to become an international reference for artificial intelligence. An integrated approach to artificial intelligence.

The PRAIRIE Institute aims to catalyze exchanges between the academic and industrial worlds, to train the new generations of researchers in artificial intelligence, and to play a role in the animation of the community. Transfer and innovation will be part of its mission, together with scientific advances. The works will highlight an integrated approach of the following two traditional axes of research:

– upstream research, backed by partner institutions of excellence in France and abroad;

– business-oriented research and applications, backed by industrial partners who are often also world leaders in their fields.

Integrated research themes will facilitate synergies between the two axes of the Institut PRAIRIE and allow researchers to easily move from one to the other.

4.7.4   Google and the École Polytechnique (X) Launch a Chair in Artificial Intelligence

On March 29, 2018, the École Polytechnique and Google France announced the creation of an international chair of teaching and research in artificial intelligence entitled "Artificial Intelligence and Visual Computing". The objective is to strengthen the ecosystem of excellence in AI by attracting international talents and training tomorrow's researchers and innovators. This new chair will be managed by Marie-Paule Cani, a professor at the École Polytechnique, and funded by Google France to support the training of a new generation of talents in artificial intelligence. This support is multiform, and targets training, research, as well as international outreach. As far as training is concerned, the new Master's degree program "Artificial Intelligence and Advanced Visual Computing" of the École Polytechnique, proposed in association with Inria, ENSTA ParisTech and Télécom ParisTech, will benefit from the support of the Chair as soon as the reception of its first promotion in September 2018. The Chair will participate in the scholarship program currently being developed to provide access to the program to all promising candidates. Google France will also offer internships for X students as part of its research internship programs. Finally, awareness actions (AI seminar: ethics and law, round table discussions, workshops) will be conducted to allow students to picture themselves in artificial intelligence jobs and research.

4.7.5 Samsung Is Going to Create a Research Center Dedicated to AI in France

Emmanuel Macron received the strategy director of Samsung. He is to announce measures aimed at developing this technology.

The Élysée's charm offensive to win over AI specialists is bearing fruit. Samsung is going to create a research center in France dedicated to this technology that has already started shaking many sectors of the economy. The South Korean group will build its third largest global research center, after the South Korea and United States ones. It will be directed by Luc Julia.

4.7.6 Higher Education Delves into the Artificial Intelligence "Black Box"

Personalization of learning help with course choices. Artificial intelligence is taking over higher education. Between technological advances and ethical issues, the subject feeds reflections and debates.

Let us go back to the EducPros conference, which took place on March 29, 2018 in Sophia-Antipolis, in partnership with Educazur and Inria.

Cognitive sciences, neurosciences, artificial intelligence. In recent years, a whole scientific vocabulary has emerged in the education sector. Researchers and entrepreneurs are redoubling their energy to design and build new tools, with one goal: improve courses and course guidance for young people.

Evidence indeed that the subject is strategic for the sector, the French AI strategy, unveiled by Emmanuel Macron on 29 March, 2018, is substantially targeted to higher education and research.

Promising new educational tools [2] Finally, they think that it would be a shame if we developed a technophobic approach of education, which would look at the progress of artificial intelligence solely as a threat. Technology is just a means to an end, it is up to us to put it at our service!

Through their consultant's profession [2], they see tools with incredible educational opportunities emerge in the still little explored field of artificial intelligence.

So far, there has been a lot of interest in Massive Online Open Courses (MOOCs) as a new way for the future, with digital technology opening massive student access to courses. But artificial intelligence is already taking us further: tomorrow, each teacher might be equipped with portable artificial intelligence able to multiply his/her personalized presence with many students, to follow what is mastered and what is not, and offer each student tailored tutorship. We would enter the era of "Massive Individual Personalized Education"! And we would finally go back to the foundation of the teacher–student relationship as Plato theorized it through the figure of Socrates, based on interpersonal dialogue and transmission.

This future is not distant, as these changes could be experienced by the next generation of students and schoolchildren. And yet, in Europe in particular, nobody talks about it. But as always in the history of humanity, it will be those who have learned how to make the most of available technologies that will be best endowed. This is the major educational challenge of the twenty-first century, it must be grasped right now.

4.7.7   French Start-Ups, Ready for the Artificial Intelligence Battle

In the wake of the Villani report on AI, "France Digitale" mapped a growing ecosystem. It listed 271 nuggets on the territory, and fundraisers that broke a glass ceiling. "Today, a start-up that raises 2 million euros is not even noticed anymore". In the world of artificial intelligence, the financing of start-ups has taken a new dimension, notes David Bessis, co-founder of Tinyclues and member of France board. The association that brings together entrepreneurs and investors hosts 'France is AI', which has just issued a map of the ecosystem of French AI start-ups.

4.7.8   Comments [3]

The current AI behemoths (the United States, China) and the emerging countries in the discipline (Israel, Canada, the United Kingdom) have developed based on sometimes radically different models.

France and Europe will have to invent a specific model to dig in their way through the AI world stage.

Hence an ecosystem of data. Hence the need to increase the visibility of those who make AI.

Focus on four strategic areas, namely: healthcare, the environment, transport-mobility, and defense-security.

Set up sectoral mutualization platforms.

Need to increase the means of calculation for research and make careers more attractive in public research.

Research is at the front of the world stage as regards mathematics and AI researchers, but it has difficulties in implementing industrial or economic applications. It suffers from the brain drain to the American behemoths. Hence the idea to create interdisciplinary AI institutes.

It is necessary to augment the calculation power available to research and to make careers in the public sector more attractive.

The world of work is at the dawn of a great transformation and is still poorly prepared to it. The uncertainties about the consequences of the combined development of artificial intelligence, automation, and robotics are huge, especially as regards the jobs to be destroyed or created.

If it is assumed that, for most professions, individuals will work in tandem with a machine, it is necessary to define non-alienating complementarity that will develop specifically human capabilities (creativity, manual dexterity, problem-solving ability).

Create a public hall of work transformation.

In this respect, experiments could be carried out to build devices targeting certain populations of individuals, whose jobs are considered to be the most at risk of automation and for whom it will be difficult to start a professional transition alone. It is therefore a question of partly breaking up with the sole logic of individual responsibility in one's own professional transition.

"My opinion is that the state is currently not ready to take responsibility for the technological transition and the consequences it will have on workers' employment".

How many jobs have been created by AI in France, in Europe, over four years, for example?"

Try new ways of financing vocational training to take value transfers into account.

Train AI talents at all levels. A clear objective must be set, i.e., to increase threefold the number of people trained in artificial intelligence in France.

Artificial intelligence at the service of a greener economy.

Research on the optimization of energy resources: the idea will be to support projects at the crossroads of life sciences, ecology, and climate and weather research.

The consumer must be a player in the "greening" of these technologies.

"My opinion is that when I see how hard it is now to get out of a productivist and polluting agriculture, will automation from AI improve the situation or make it worse?"

Think of a greener AI.

We must think of breakthrough innovation in the field of semiconductors.

What ethics for AI?

The recent advances of AI in many areas (autonomous cars, image recognition, virtual assistants) and its growing influence on our lives are consolidating its place in the public debate.

Open the black boxes.

Many of the current ethical issues are due to the opacity of these technologies.

Think ethical as early as conception.

Beyond engineers' training, ethical considerations should irrigate the very development of artificial intelligence algorithms.

More generally, the increasing use of AI in certain sensitive fields such as the police, banking, insurance, justice, or the army (with the issue of autonomous weapons) calls for a real social debate and a reflection on the question of human responsibility.

Create an AI ethics committee.

For an inclusive, diverse AI.

Mixity and diversity: acting in favor of equality.

Develop digital mediation and social innovation for AI to benefit to everyone.

In order to redistribute these innovative capacities, the public authorities could launch specific programs to support social AI innovation and equip social actors so that they can benefit from AI advances.

"My opinion is that these comments appear as wishful thinking, as if we were not in a world dominated by GAFA, which already hold the AI world in their hands. I am afraid we are already off side and are seeking to limit the damage as regards GAFA, China and the United States."

4.7.9   Google Publishes a List of Seven Ethical Principles in Artificial Intelligence

https://www.frenchweb.fr/google-publie-une-liste-de-sept-principes-ethiques-en-matiere-dintelligence-artificielle/327614.

Following the Maven scandal, Google CEO Sundar Pichai released the company's 7 principles for artificial intelligence.

As a reminder, the Maven program refers to a contract that had been signed between Google and the Pentagon. As part of this project, the artificial intelligence technologies (video image interpretation) of the Mountain View firm were to help the US military improve visual recognition and information gathering by its drones (non-armed aircraft) and even surveillance and strikes per drone. The discovery of this agreement aroused an outcry among employees, and caused more than a dozen resignations. A petition signed by more than 3,000 employees was also launched, damaging the image of Google in the process. Ninety academic personalities also wrote a letter asking Google to break the contract and commit not to develop military technologies. Finally, Diane Greene, head of Google Cloud, announced to her teams that the contract, which runs until 2019, would not be renewed.

With this post listing the principles of Google regarding artificial intelligence, the company seems to want to show a clean record and prove that it does not take this subject lightly.

For example, socially beneficial projects fight against cognitive biases [1].

The first principle is that technology will have to be used for "socially beneficial" projects. "When we look into the opportunities for the development and use of AI technologies, we will consider a wide range of social and economic factors, and we will pursue when we estimate that the likely overall benefits far outweigh the foreseeable risks and drawbacks. foreseeable disadvantages". It will also be necessary for it not to create or strengthen any bias.

It will also be necessary for the application to abide by security measures. "In appropriate cases, we will test AI technologies in constrained environments and monitor their functioning once they have been deployed." The application will also have to be "responsible" to the population. "Our AI technologies will be under appropriate human direction and control," said Sundar Pichai. AI technologies will have to abide by private life protection measures, including asking for the consent of the persons concerned.

In its penultimate principle, Google wants to "maintain high standards of scientific excellence." "We will work with a range of stakeholders to promote thoughtful leadership in this field, by relying on scientifically rigorous and multidisciplinary approaches." And finally, Google will have to ensure that the implemented technologies cannot be "diverted" and contravene the principles listed above. "Many technologies have multiple uses. We will strive to limit potentially harmful or abusive applications."

Even if these principles are paved with good intentions, they remain rather vague. Moreover, in the same post, Sundar Pichai explains that Google intends to keep on working with the government and the military in cybersecurity, training, recruitment, research, or rescue. He says, "we want to make it clear that we do not develop AI for weapons". But finally, if we look at the Maven program that concerned unarmed devices, did it not abide by the principles listed above? With this list, Google finally does not seem to have had many doors closed. The agreement between the Pentagon and Google on the Maven project amounted to $ 9 million.

Having long been a subject of study and controversy, philosophical intelligence is often associated with a person's reasoning and thinking skills. Historically, in its manifestation, it is the opposite of instinct, which is thought to correspond more to a reflex than to an elaborate thought. The notion has evolved over time, according to discoveries, eras, and thinkers.

4.7.10 Psychological Intelligence

In 1905, the French government asked psychologist Alfred Binet to establish a tool for measuring human intelligence. This was how the metric scale of intelligence, at the origin of the famous intelligence quotient (IQ) test or its derivatives was born.

These tests, although questionable, are supposed to allow for comparisons between people's intellectual performances through various and varied exercises, in a maximum of domains such as logic, reasoning, memory, or emotion. The individual is evaluated as compared to the average score of a group of representative persons.

Some specialists even speak of plant intelligence, as some of them are able to communicate or to distinguish the self from the nonself.

References

1. M. Agogué, *L'innovation orpheline: lutter contre les biais cognitifs dans les dynamiques indus-trielles*. (Presses des Mines, 2013)
2. C. Edouard, Bouée and François Roche. *La chute de l'Empire humain: Mémoires d'un robot*. (Grasset, 2017)
3. C. Villani, *Donner un sens à l'intelligence artificielle* (2018)

Chapter 5
Plant Intelligence

Abstract Stefano Mancuso and Alessandra Viola in their book "Brilliant Green" present a new paradigm of our understanding of the plant world. Mancuso declared. "Plants are made of a large number of basic modules that interact like nodes in a network".

Reference book: Brilliant Green: The Surprizing History and Science of Plant Intelligence, by Stefano Mancuso, Alessandra Viola [2].

Are plants smart? Yes, they are, and much more so than we could imagine, Stefano Mancuso answers. A world-renowned scientist and the founder of plant neurobiology, he is the first to have demonstrated that, like all living beings, plants discern shapes and colors, memorize data, and communicate. They have a personality and develop a form of social life based on mutual help and exchange.

A true ecological manifesto, this pioneering book [2], which received international recognition, plunges us into an incredible journey into the core of the plant world. A world that is currently forming more than 99% of the biomass, and proves essential for humanity. Because if plants can live very well without us, we would not survive long without them!

At a time when we are seeking new modes of life, when natural resources are being depleted, everything is to be learned from the plant kingdom, and human survival and future depend on it.

Can they communicate with one another, solve problems, and sail through their environment?

Or are they inert, incapable of a collective behavior?

Philosophers and scientists have pondered these questions since the days of ancient Greece, and have most often concluded that plants represent a "lower degree of life": indeed, they are too silent and sedentary, far too different from us.

However, discoveries over the last 50 years have invalidated these ideas, shedding new light on their extraordinary abilities and the complexity of their inner lives.

In "Brilliant Green", Stefano Mancuso and Alessandra Viola present a new paradigm of our understanding of the plant world. Because of our cultural prejudices and arrogance, we underestimate plants.

In fact, they process information, sleep, tell each other, and point out to each other that far from being passive machines, plants are intelligent and conscious.

C. Lexcellent, *Artificial Intelligence versus Human Intelligence*, SpringerBriefs in Applied Sciences and Technology, https://doi.org/10.1007/978-3-030-21445-6_5

Plants have a lot to teach us: from network building to innovations in robotic and synthetic materials, but not just that, we now better understand how they live.

Partly through the botany lesson it teaches, partly through its "manifesto" aspect, "Brilliant Green" [2] is an engaging and passionate study of the inner functioning of the plant kingdom.

Plants make up ninety-nine percent of the Earth's biomass with reference to humans and animals that represent only a trace.

Plants found an excellent advocate in Charles Darwin who wrote "It has always pleased me to exalt plants as organized beings".

"I dream of a botany that could determine itself on its own, according to its own rules, no longer lagging behind animal or human physiology: taking the plant itself into account, as an original form of life, as a model of autonomy and restoration of the environment, it could find its place in the center of life sciences. In our world of hard cash, of showing off, advertising, noise, pollution and brutality, what better testimony than that of plants, beautiful and useful, discreet and autonomous, silent and completely non-violent?", as Francis Hallé stated [1].

Now, the first question to be asked is: "Is the brain really the only production site of 'intelligence'? Is a brain without a body still intelligent, or on the contrary, will it appear only as a group of cells without any special characteristics? Will we be able to find any trace of intelligence? The answer is unequivocally no. The brain of our greatest genius is not smarter than our stomach. It is not a magical organ and it certainly cannot create anything on its own. Any information coming from the rest of the body is fundamental to provide a clever answer.

Well, in plants, cognitive and bodily functions are not separated but are present in every cell; a real, living example of what artificial intelligence scientists have called an "incarnate agent," that is, a smart agent that interacts with the world with its own physical body.

We emphasized that evolution had given plants a modular structure, i.e., did not concentrate functions in individual organs but distributed them across their entire living space. Therefore a plant has no lungs, liver, stomach, pancreas, or kidneys. So, why does the absence of a brain prevent them from being intelligent?

Let us look at a root, the plant part where Darwin perceived the seat of decision and the management of capabilities.

The root tip directs growth and explores the soil for water, oxygen, and nutrients (Figs. 5.1, 5.2, 5.3).

Each plant has millions of root tips; the root system of a very small plant can have more than 15 million of them.

Each root tip continuously detects many parameters such as volumetric mass density, temperature, humidity, the electric field, light, pressure, chemical gradients, the presence of toxicants (poisons, heavy metals), vibrations, the presence or absence of oxygen and carbon dioxide, etc.

No automatic response could meet the previously detailed root tip "specifications".

These roots do not work alone but along with millions of others, to make up a plant root system.

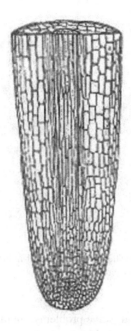

Fig. 5.1 The tip of the root: each root tip is a sophisticated memory organ

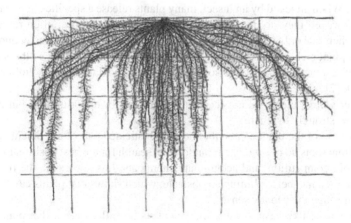

Fig. 5.2 Root system of an 8-week-old maize plant. Such a system is made up of 10 million root tips

Plants are smart. Plants deserve rights. Plants are like the Internet—or more exactly the Internet is like plants.

"A brain without a body produces the same amount of intelligence as the hazelnut it looks like" says Mancuso [2].

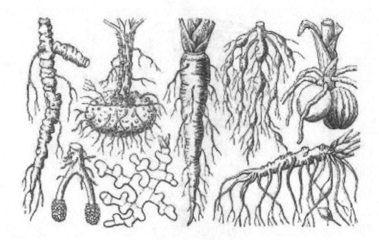

Fig. 5.3 Examples of root systems. The roots are the half-hidden parts of plants and the most interesting one. The illustration shows various root types

"Intelligence is the capacity to solve problems; funnily enough, plants are good at solving their own problems," adds Mancuso.

Finally, plants have developed an incredible array of toxic compounds to repel predators. When attacked by an insect, many plants release a specific chemical compound. However, they do not just throw compounds around, but often release the precious chemical only in the attacked leaf. Plants are both cunning and economical.

"Each plant's choice is based on this type of assessment: what is the smallest amount of resources to be used to solve the problem?", Mancuso and Viola write in their work [2].

In other words, plants do not react only to threats or opportunities, but have to decide how strongly to react.

The bottom of the plant can be the most sophisticated part of all. Scientists have observed that roots do not wander at random but search for the best position to absorb water, avoid competition, and accumulate chemicals. In some cases, the roots will change their course before hitting an obstacle, which shows that plants can "see" an obstacle through their many senses.

Humans have five basic senses. But scientists have discovered that plants have at least 20 different senses to monitor the complex conditions of their environment. According to Mancuso, their senses roughly correspond to our five senses, but they have additional senses that measure humidity, detect gravity, and detect electromagnetic fields.

Plants are also complex communicators. Today, scientists know that plants communicate in a variety of ways. The best known of these is chemical volatile substances—why some plants smell so good and others so bad—but scientists have also discovered that plants also communicate via electrical signals and even vibrations.

"Plants are wonderful communicators: they share a lot of information with neighboring plants or other organisms like insects or other animals." The smell of a rose, or something less fascinating like the stink of rotten meat emitted by certain flowers, is a message to pollinators."

Many plants will even warn other plants of the same species when danger is close. If a plant is attacked by an insect, it will send a chemical signal to its companions as if to say, "Hey, I'm being eaten—so get prepared to defend yourselves." Researchers have even discovered that plants recognize their relatives, by reacting differently to plants from the same parent than to those from a different parent.

"In recent decades, science has shown that plants are sentient, weave complex social relationships, and can communicate with one another and with animals," Mancuso and Viola write [2].

And it turns out that Darwin was probably right from the beginning. Mancuso found growing evidence that the key to plant intelligence lies in the root or the root of the root. Mancuso and his colleagues recorded the same signals from this part of the plant as those from neurons in the animal brain. One root apex may not be able to do much. But instead of having a single root, most plants have millions of individual roots, each with a single radicle.

Thus, instead of a single powerful brain, Mancuso argues that plants have a million tiny computer structures that work together within a complex network, which he compares to the Internet. The strength of this evolutionary choice is that it allows a plant to survive even after losing 90% or more of its biomass.

"The main driver of plant evolution was to survive the massive removal of a part of the body," Mancuso declared. "Plants are made of a large number of basic modules that interact like nodes in a network. As plants do not have single organs or centralized functions, they can tolerate predation without losing their functionality. The Internet was born for the same reason, and inevitably reached the same solution."

Do plants sleep? Examining their leaves in the diurnal and nocturnal phases is enough to be convinced.

In the lyric genre, Jagadish Chandra Bose (1858–1937) was an Indian scientist and a supporter of the fundamental identity between plants and animals "Trees have a life of their own like us, they eat and grow ... face poverty, sorrows and suffering. This poverty can lead them to steal, but they also help one another, develop friendships, sacrifice their lives for their children."

References

1. F. Hallé, *Eloge de la plante, Pour une nouvelle biologie* (Poche, 2014)
2. S. Mancuso, A. Viola, *Brilliant Green: The Surprizing History and Science of Plant Intelligence* (Island Press, 2015)

Chapter 6
Animal Intelligence

Abstract Emmanuelle Pouydebat in her book "Animal Intelligence" presents the amazing capabilities of primates, birds, insects, or fish. Thus, she does not believe in the existence of a clearcut boundary between intelligence and non-intelligence, or that intelligence is the prerogative of human beings.

Emmanuelle Pouydebat, a biologist at the National Museum of Natural History and at the CNRS, is a specialist in the handling of tools by animals. In her book "Animal Intelligence" [3], she presents the amazing capabilities of primates, birds, insects, or fish. According to her, intelligence could in fact be multiform, and not meet the classic definition that intelligence is linked to the use of tools or to the presence of motor skills, such as bipedalism or the presence of an opposable thumb. Thus, she does not believe in the existence of a clearcut boundary between intelligence and non-intelligence, or that intelligence is the prerogative of human beings.

For example, ants are able to zigzag 600 m without using tools to search for food, and then go back in a straight line [3], thanks to an internal pedometer system. Intelligence can also be evidenced by ingenuity, despite the lack of an apparent skill because of claws, or the lack of an opposable thumb. Rats, for example, succeed in piling up objects to escape from an enclosure, otters and raccoons manage to open boxes with their claws.

Additionally, some animals use objects to achieve their ends. Otters are able to fetch a pebble and throw it to break a shell. As for the bottlenose dolphin, it uses plant sponges to protect its nose when it forages for food in the sedimentary bottoms.

Some animals are also able to get organized and cooperate to achieve a goal. Among ants, for example, some individuals bring leaves together while others "sew" them together, sticking them together with silk produced by the larvae.

"Therefore everyone has a different intelligence according to the environment, and we cannot classify species on a one-dimensional intelligence scale ...".

Human beings have a place in this whole, but it is "only a drop of water" in the animal kingdom.

The more interested humans are in animal intelligence, the more they need to reassure themselves about their own intelligence. If crows know how to make tools, ants

C. Lexcellent, *Artificial Intelligence versus Human Intelligence*, SpringerBriefs in Applied Sciences and Technology, https://doi.org/10.1007/978-3-030-21445-6_6

go to war like nobody, elephants count without making a mistake, and chimpanzees paint pictures, what is left to humans?

Certainly, the need to wonder about themselves and the world. Well, let us use that faculty to try to find a connection between what is called intelligence and an anatomical feature of the brain.

If we are indeed the most intelligent species on the planet, it must be visible somewhere in the organ we are so proud of.

Thus, we can consider the "brain mass versus body mass" criterion, which could reassure us, a kind of linear curve between the shrew (the "most stupid" one) and the sperm whale ("the most intelligent" one) [1]. The curve allows them to deduce "the neural factors that determine intelligence" !

Finally, the main contribution of the ethologist Frans de Waal [2] is the highlighting of the phenomenon of reconciliation in many primates species after a conflictual interaction, an ability that was previously considered as typical of the human species. He wonders what ultimately makes the difference between humans and bonobos or chimpanzees, with which we share 98.7% of our genes.

He stresses the difference between communities of chimpanzees on the one hand, i.e., brawlers capable of pernicious coalitions to impose the choice of the colony leader, and bonobos, who are pacifists and whose females regulate conflicts between males by offering a sexual intercourse, of which bonobos are reputedly insatiable. Is there not a behavioral intelligence in bonobos, especially in females?

References

1. U. Dicke, G. Roth, Neuronal factors determining high intelligence. Phil. Trans. R. Soc. B **371** (2016)
2. F. de Waal, *Sommes-nous trop << bêtes >> pour comprendre l'intelligence des animaux ?* [<< *Are We Smart Enough to Know How Smart Animals Are?* >>],. Éditions Les Liens qui libérent (2016)
3. E. Pouydebat, *L'intelligence animale, cervelle d'oiseaux et mémoire d'éléphants* (Odile Jacob, 2017)

Chapter 7
Intelligence

Abstract It is very difficult to give a clear definition of human intelligence. One can only speak about the multiple forms of intelligence. The chapter ends with some reflections of philosopher Emmanuel Fournier.

7.1 Is Human Intelligence Really at Risk?

Let us recall that access to symbolic systems is part of human intelligence.

Measuring human intelligence is an obsession materialized by IQ tests (see Laurent Alexandre's words on this subject [1]).

The fluctuations of the average IQ score have been monitored over the years. The Flynn effect corresponded to its increase [6] steadily recorded in the developed countries.

But since 1995, there has been a decline in the overall IQ level in developed countries. For example, "between 1999 and 2009, the average IQ of French people has dropped by four points" [4].

This overall IQ decline, observed in several studies, raises many questions. What could be the reason for it? The environment? A genetic factor? This brings us back to the question of knowing what is innate and what is acquired. Is intelligence "a gift of nature", or is it acquired in a given social, cultural context?

"In short, it is a question of knowing whether intelligence is a natural datum, whether individual (the gift theory) or collective (the racialist concept). On the contrary, it can be the result of a social construction, which offers everybody the hope of a good intellectual development if conditions are met."

Let us go back for a moment to what is measured during an IQ test.

Confusion between performance and competence Like any test, the IQ test can be challenged. Either we are adapted to the test thanks to the richness of our previous life, or we are not.

The IQ is a game, and as in any game, those who know the rule best play it best. The IQ measures the level of an individual versus the level of a group in several

C. Lexcellent, *Artificial Intelligence versus Human Intelligence*, SpringerBriefs in Applied Sciences and Technology, https://doi.org/10.1007/978-3-030-21445-6_7

domains, including verbal and numerical domains. But the IQ does not evaluate the "intellectual weight" of an individual.

IQ test results show performances of a group of individuals at time t. And we must not confuse performance and competence. Performance, as in the case of a sportsman, for example, may be more or less satisfactory depending on when the test is done. For example, if a person suddenly spends a lot of time in front of a screen instead of simply reading, we can expect his/her writing performance to decrease.

We will see other concepts and definitions of intelligence later on, including the eight forms of intelligence as defined by Howard Gardner [8].

Intelligence as a power The IQ test is believed to only measure "facets of intelligence". In the book "The Human Brain" [14] prefaced by Edgar Morin and Massimo Piattelli-Palmarini, a thesis proposes that intelligence is not a thing, nor an object or an organ. It would rather be a potential, a power to do something, to learn to do something.

As François Jacob writes in "Logic of the Living" [10], "Anthropological attributes are rigorously prescribed by the genetic program. But they only determine potentialities—the power to walk, to speak any language, to understand ..."

Intelligence is the attribute that expresses itself in the thinking power. It only wears off if you do not use it! It is like "freedom of the press", the motto of the newspaper "Le canard enchaîné".

Available technologies do not necessarily make us smarter [2] Several explanatory hypotheses are being explored, including the impact of pollution and endocrine disruptors on the functioning of our brains.

A decrease of the IQ, an increasing number of children with hyperactivity or learning disabilities: the most serious tests reveal what seemed unimaginable 20 years ago, i.e., the decline of human intellectual abilities. Have we entered a kind of "reverse evolution"? The question is posed by eminent researchers. In the dock are endocrine disruptors, which have invaded our lives and threaten babies' brains. Revelations about a disturbing phenomenon are presented in a documentary by Sylvie Guilmain and Thierry De Lestrade (DEMAIN, TOUS CRÉTINS?, Coproduction CNRS Arte 2017).

Still, this result is quite counterintuitive for everyone. How many times have we told ourselves that our intellectual abilities and our efficiency were strongly increased by technological progress? That we had access to lots of information that in fact made us smarter? And yet, such is not the case. Richard Flynn even considers that the downward trend of the IQ could be sustained because the maximum IQ threshold is thought to have been reached in the nineteenth and twentieth centuries in the wake of the industrial revolutions and the dramatic improvement of living conditions, and access to education of populations that had previously remained remote.

It is therefore not technological progress that mechanically increases the population's intelligence level, but the policies implemented to adapt the population to a new technological context. In an essay entitled "The Race Between Education and Technology", [9] Claudia Goldin and Lawrence F. Katz show how the nineteenth and twentieth centuries were periods of intense state mobilization to educate and train populations that needed to adapt to the new demands of the productive system. But it is high time we started a new page in educational history. Because we are at the dawn of a new wave of major technological disruption that is giving rise to new, presently unsupported educational needs.

New educational needs What does the work of a legal assistant in a company look like when he/she has at his/her disposal high-performance software for jurisprudence research? Similarly, should the skills of a worker in the automobile sector remain the same when his/her job is to interact with a "cobot", an intelligent and collaborative robot? In both these examples, not only are the required technical skills no longer the same, but neither are social skills: intellectual agility, the decision-making ability, creativity, etc. become much more important.

Business needs are changing at a high speed, which entails that the training requirements are also changing. However, only few companies wonder about the skills they will need over the next 10 years, or those they will no longer need.

Europe is well placed Set up an observatory of future professions involving social partners in all European countries to monitor which jobs are emerging, grasp the evolution of the skills needed for different job categories, appears as a simple and essential idea, which will then make it possible to change training devices, which we know are not up to the challenges.

We have many reasons to remain optimistic, especially in Europe. Our training systems are recognized worldwide, especially in sciences and mathematics. It is often said that the French mathematician and the German engineer form a winning duo at the time of artificial intelligence! But we must think of ways to lead the whole society—and not just the most qualified workers—to a training dynamic adapted to tomorrow's economic and technological challenges.

7.2   Definitions of Human Intelligence

Mickael Klopfenstein proposes definitions of intelligence such as "the autonomy to produce an abstract treatment of data for operational efficiency" or "the ability of a system to perceive an organized sense within a global reality and to react following an analysis of perceptions with a view to a purpose preformed by the system [11].

According to Daniel Andler: "intelligence is what one expects from the other".

In his book "The Element," Ken Robinson [15] says that he usually asks the participants to his lectures to rate their intelligence. He asks them this question: "Where do you rank your intelligence on a 1-to-10 scale?"

He is always surprised to see people answer this question (the majority rank themselves between 8 and 6), whereas according to him this question makes no sense. People should answer that the question is not good and should be replaced by this one:

What form of intelligence is yours? In what way are you smart?

This question suggests that there are various ways of being smart.

Ken Robinson recalls that scientists never managed to agree on a single and common definition of intelligence. In "The intelligence of the heart" [5], Isabelle Filliozat adds that IQ tests only evaluate the mastering of language and mathematical logic. The school system emphasizes these two mental faculties and it is in this sense that IQ tests have a "predictive" value of academic success. In the common spirit, intelligence has become synonymous with the ability to answer a verbal and logico-mathematical test. Yet, the IQ does not measure the different forms of intelligence.

Intelligence is not synonymous with academic ability.

7.2.1 The Multiple Forms of Intelligence

The American psychologist Howard Gardner defined eight types of intelligence [8] (taking care not to declare this list as firm and definitive). He defined intelligence as:

"The ability to solve problems or produce valuable goods in a specific cultural or collective context. The problems to be solved range from inventing the end of a story to anticipating a mat at chess, via mending a quilt. Goods range from scientific theories to musical compositions, via victorious political campaigns."

Howard Gardner recognized the verbal and logico-mathematical forms of intelligence assessed by IQ tests, but placed them on par with others. Education should treat each type of intelligence equally so that all children have the opportunity to develop their own abilities and passions.

– Language intelligence

Linguistic verbal intelligence is the ability to use language to express and understand ideas. This sensitivity to words and language, oral exchanges, reading and writing, is useful to writers, but also to politicians or teachers.

– Musical intelligence

Musical intelligence relies on ear acuity and perception of rhythm. It can be useful to a mechanic who will identify a problem in a car engine just by hearing the noise it makes!

– Logico-mathematical intelligence

Logico-mathematical intelligence is the ability to hold a logical reasoning, to calculate, to analyze.

– Spatial intelligence

Spatial intelligence is the ability to represent oneself in three dimensions, to orient oneself, to have a good sense of orientation. It is particularly useful to sailors, engineers, surgeons, sculptors, painters, or architects.

For example, this form of intelligence is developed by video games.

– Kinesthetic intelligence

Kinesthetic intelligence is the intelligence of the body, the ability to use the body to express emotions, to memorize information, to master a gesture, to create. Let us think of dancers and athletes, but also surgeons and craftsmen.

– Interpersonal intelligence

Interpersonal intelligence defines our relationship with others: the ability to understand and work with others. It is being smart with others!

– Intrapersonal intelligence

Intrapersonal intelligence is the knowledge of oneself. Gardner defines it as "the faculty to form a precise and faithful representation of oneself and use it effectively in life."

– Naturalistic intelligence

Naturalistic intelligence is the ability to observe and respect nature, to recognize, identify and classify animals, plants, and minerals.

7.2.2 Analytical, Creative, and Practical Forms of Intelligence

Robert Sternberg, a professor of psychology, defined three types of intelligence and recognized that they were complementary, and were found at various degrees and integrations in individuals.

– Analytical intelligence

It is the ability to solve problems academically and pass traditional IQ tests. These types of school tasks have only one correct answer. People whose predominant form of intelligence is analytical are reluctant to come up with new ideas on their own.

– Creative intelligence

It is the ability to face new situations and find original solutions. Creative intelligence synthesizes existing knowledge and skills and paves the way for divergent answers. People with a predominantly creative form of intelligence tend to rely on their intuition but may score low at standard IQ tests because these do not sufficiently measure creativity and intuition. People with a dominant creative intelligence are most likely to innovate, think differently, and think "off the tracks".

– Practical intelligence

It is the ability to solve everyday life problems and challenges. This form of intelligence is practical in that it helps to understand what needs to be done in a given situation and to do it. For Sternberg, practical intelligence consists in applying the analytical and creative forms of intelligence to everyday situations by self-adapting or by reshaping the environment.

7.2.3 Emotional and Social Intelligence

Daniel Golemean, a psychologist and author of the bestseller "Emotional Intelligence", distinguishes emotional intelligence from social intelligence. They both allow us to live in harmony with ourselves and the world around us.

– Social intelligence

We are social beings: we are designed in such a way that we need contact, talking with others, touching and being physically touched. Neurosciences show that we exist with other people, depending on them, thanks to them, relatively to them. Golemean integrates understanding others and facilitating relationships between people in the notion of social intelligence.

– Emotional intelligence

According to Golemean, emotional intelligence includes competences such as:

– the ability to remain motivated and persevere despite adversity and frustration,

– the control of one's instincts,

– the ability to postpone a satisfaction,

– the ability to regulate one's mood and to prevent distress, to alter reasoning skills,

– empathy,

– hope.

7.2.4 Heart Brain and Visceral Brain

According to Robert Cooper, intelligence operates through three brains:

– the brain in the skull,

– the brain in the heart,

Intelligence of the heart, dear to Isabelle Filliozat [5], gives access to deep motivations, to true emotions in oneself and in others, to autonomy. According to Isabelle Filliozat, the autonomous individual is the one who forges his/her own rules of life, the one who listens to his/her heart.

"The intelligence of the heart is based on the giving, receiving, asking, and refusing competences. It also requires us to know how to listen to others, decode their messages and settle our conflicts nonviolently."

– the visceral brain.

Neurons are present in our intestines. The belly is endowed with an intelligence of its own, although not conscious. But "it is above all unconscious processes that control our lives—because consciousness always lags half a second behind our reality" (source: The intelligence of the Belly). The belly is then believed to be the "manager of our instinct, provider of unconscious information, and regulator of our scruples". That is why we talk about "visceral reactions".

"Intelligence can manifest itself under forms that have little or nothing to do with numbers and words. We reflect on the world in all the ways we experience it, including the different ways we use our senses. [15]

This multiplicity of the forms of intelligence makes me think of a patchwork where everything is in everything and vice versa!

I am now going to provide a more classic definition of human intelligence.

7.3 A More Classic Definition

Short note: in the dairy industry in the Franche Comté region, robots are prohibited for milking in the manufacturing process of PDO (European standard) comté cheese.

Would that mean that human intelligence for once won against AI?

Intelligence commonly refers to the potential of an individual's—animal or human—mental and cognitive abilities to solve a problem or adapt to the surrounding environment. It is often limited to the brain. It can be subdivided into different components: we speak of practical intelligence, collective intelligence, emotional intelligence, or business intelligence for example. By extension, it was adapted to machines and called artificial intelligence.

However, the concept of intelligence is extremely complex and debated. The term is defined differently depending on the field in which it is treated.

Intelligence in philosophy Long the subject of study and controversy, intelligence in philosophy is often associated with a person's reasoning and thinking skills. Historically, it is opposed to instinct in its expression, as instinct corresponds more to a reflex than an elaborate thought. Its notion evolved over time according to discoveries, times, and thinkers.

Intelligence in psychology In 1905, the French government asked psychologist Alfred Binet to establish a tool for measuring human intelligence. Thus was born the metric scale of intelligence, at the origin of the famous intelligence quotient (IQ) test or its derivatives.

These tests, although questionable, are supposed to allow for comparisons between people's intellectual performances through various and varied exercises, addressing a maximum number of domains such as logic, reasoning, memory, or emotion. The individual is evaluated as compared to the average score of a group of representative persons.

Cunning, practical intelligence [16] The Greeks called it metis, i.e., that particular form of intelligence that combines tactics and a spirit of subtlety. Although it is hard to define, it is present everywhere: in the minds of the strategist, the hunter, or the handyman.

According to the ancient Greeks, there is not a single god ordaining everything and having created everything. Gods are everywhere in the world. Multiple, diverse, they take all forms. Zeus is the king of the gods, the master of sovereignty in all its appearances. Goddess Metis was his first wife. Hardly was she pregnant with the one who was to become Athena, the goddess of wisdom and intelligence, than Zeus swallowed her, relegating the cunning to the depths of his own belly and giving birth to his own daughter, thus integrating intelligence and cunning, i.e., metis.

Metis refers to that ability of intelligence that does not correspond to abstraction, but to practical efficiency, to the field of action, to all these useful skills, to the craftsman's skill in his profession, his handiness, magical tricks, war tricks, deceits, dodges, and wily deeds of all kinds.

In any conflict or competition situation, victory can be obtained in two ways. Either because one is the strongest on the ground in question, or by using processes aimed at distorting the trial and making the one expected to be beaten triumph. The metis can be considered as what brings about fraud, or on the contrary what brings surprise and the revenge of the weakest. In one case, it puts on a face of lies, deceit; in the other case, it is the absolute weapon, which secures victory over others in all circumstances. The second characteristic of the metis is that it is always performed in uncertain, ambiguous situations. For example, if two men are fighting, everything can tip over one way or the other at any moment, but, during the trial, the man with the metis is the one who will be able to show premeditation and vigilance. The metis is the lookout, the man who spies to hit the opponent at the most unexpected moment. Spying, in Greek, is a term that is used indistinctly for fishing, hunting, and warfare. In French and in other languages too. A third characteristic that Homer gives to the metis is that it is always multiple, as Ulysses is. It is like the shimmering drawing of a fabric, the speckled, shining back of the snake.

The last characteristic of the metis is that it is par excellence the power of cunning, the one who acts under the cover of a mask. With it, reality and make-believe duplicate and confront each other as two opposite forms to create a deceitful illusion. And the most cunning of all is Ulysses, the master of words who, every time he speaks, pretends he is incapable of uttering a word.

Cunning is the practical intelligence of the sailor, the weaver, the carpenter, the lumberjack. It is the skill of the politician, the doctor, and the strategist. For each of these people, cunning consists of tracking the favorable circumstance, or even creating it. No doubt, this practical intelligence long remained in the background. Yet, Plato and Aristotle did not fail to explain its qualities. The first of these qualities consists in knowing how to relate the mobility of intelligence and the speed of action: it is the delicacy of the mind, vivacity, sharpness. Aristotle gives the example of the midwife cutting the newborn's umbilical cord. It is, he says, the accuracy of the glance, "which does not miss the goal to be reached". In the same vein, Plato refers to the skill of the archer stretching his/her bow towards the target. As far as the metis is concerned, the accuracy of the glance is as important as the agility of the mind. "To target" and"to conjecture" converge in Greek towards the idea of the sailor at sea or of a trip in the desert, where the paths are no longer traced and where the traveler stops and aims at a point in the distant horizon.

To find one's way in a world of moving symptoms, one needs fluid intelligence. The doctor is like a sailor holding the rudder: he/she has to guess his/her way by using all the signs that he/she can recognize and use at best. Conjectural knowledge proceeds through the detour of a comparison that makes it possible to understand an unknown event using its resemblance with a familiar event.

With cunning, we are in the presence of a real mental category, which plays on various registers. Cunning includes everything, but never "deceit" in its present common meaning. A game of the mind, of skill, of experience. Also a game of compositions that will be set up according to what is known and what is at disposal, as compared to what is seen or what can be predicted. Should we recall that we come from a Hellenic civilization, which invented drama and its superior expression, i.e.,

Fig. 7.1 Classic stress-strain
curve: strain versus
deformation; elastic "linear"
portion; "non-linear"
elastoplastic portion [12]

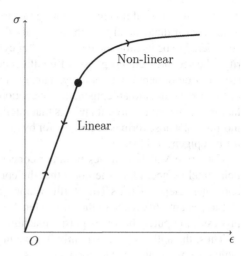

tragedy? And that in this tradition, the actor is the "hypocrite", namely the one who acts as a convincing character? In this tradition, cunning saves effort, avoids brutality. It is the trick of the oppressed against domination, the trick of the citizen against power.

Artificial intelligence Since the emergence of robotics and computer science, researchers have been trying to inject notions of human intelligence into machines. As this form of intelligence is designed and manufactured, we call it artificial intelligence.

7.4   The Metamorphoses of Philosopher Catherine Malabou's Intelligence [13]

When the author speaks of "metamorphosis", she means conceptualizing plasticity based on its pre-established human brain anchoring.

Let us note that the term "plasticity" was appropriately borrowed from the mechanics of solid materials. The behavior of materials is called "nonlinear"; if it is independent of time, it is called plastic, and if it depends on time, it is called viscoplastic [12].

As showed in Fig. 7.1, when a material is mechanically stressed, it first has a purely elastic reversible behavior, and then it "plastifies".

Neurosciences and the human sciences retained the fact that materials are malleable, and introduced the concept of "brain plasticity".

Metamorphosis corresponds to this evolution of the brain over time (growth, aging, etc.).

Catherine Malabou works to transform this concept of plasticity by confronting its relevance to the latest and upcoming computer technologies.

"It is a question of no longer wanting to purify the intelligence of its opposite, but to think of it dialectically "with its stupidity", that is to say with what philosophy considered to be its biological (the brain) or cybernetic (artificial intelligence) impurity. The second metamorphosis of intelligence results from a paradigm shift about the question of heredity in biology, namely the transition from a genetic conception of brain structures to an epigenetic conception that integrates the environment as a factor of gene expression. It appears that intelligence consists of a continuous metamorphosis, whose form is accounted for by plasticity and whose process is accounted for by epigenesis" [3].

Catherine Malabou moves from her concept of plasticity to the question of artificial intelligence. First, she notes that the epigenetic revolution has taken place in computer science: IBM's TrueNorth synaptic chips, which simulate neuronal architecture, can modify some of the parameters by themselves according to the data they process. Computers become "plastic machines" [13].

Thus, the authors [3] apparently believe in the plasticity of computer programs, which for me remains a largely open question.

Thus, the question of the form of intelligence is not worded as "what is intelligence?", but as "how is intelligence metamorphosed?".

The questionable aspect of their speech is that they put computer intelligence, for example, and human intelligence on a par.

"Stigmergy" is a fertile concept that was developed to reflect on collective intelligence.

7.5 Reflections of Philosopher Emmanuel Fournier [7]

In his "Lettre aux Ecervelés", Emmanuel Fournier thought it wise to warn us "I, a great professor, an eminent neuroscientist and a representative of the universal cognitive thought, thought it good to warn the public against the terrible danger of thinking, just as Voltaire wisely warned us against the danger of reading." He referred to "the ancient practice of thought which was perpetuated without knowing what it owed to the brain laws." "Thanks to the spectacular progress of imaging technologies and the results of the ensuing bold brain research, we finally discovered the rules that regulate thought!".

We must "learn how to watch the world progress through the prism of the brain and understand all the advantages we can draw by following the certain and undoubtable rules that our objective and new science determine day after day, in psychology, educational sciences, as well as economics, politics, or ethics.

"How did the brain rise to its throne? It owes a lot to pathology, but also to brain imaging devices."

"In order to be, do we not have to possess a brain?"

"I am told that I was pre-wired to be a thinking subject and a social being". "But I am also told that my brain preceded me and predetermines me. While our brains predestinated us to civilization, civilization shaped the material wiring through mechanisms of synaptic plasticity." Let us recall that synaptic plasticity is the property of the connections between neurons, called synapses, to change according to the use that is made of them. It encompasses the multiple mechanisms involved in the modification of synaptic transmission over time.

"This is considered as proof that the brain drives our behavior and is the cause of social life."

It is thought that "the brain is materially modified by the living experience of individuals", which seems perfectly acceptable to me.

"How can we turn thought into a science object?"

That's a dilemma, we can say that thought belongs to the realm of philosophy, perhaps does it come from the unconscious? But conceivably not from science!

Emmanuel Fournier's answer went as follows: "Considering the brain as the cause of thought leads to redefining thought as a brain product. Thought will be an ability to evaluate, calculate, reflect, or imagine ... Consciousness will be an opportunity to verbalize, describe, recognize, or plan".

Thought can be volatile, evanescent, changing, in a word, living. In short, it is not governed by theorems, Cartesian reasoning ... in short, it is not a science, and to my mind fortunately so!

It would be difficult for philosophy to stay away from a "subject" like the brain. But handing over the question of the relationships between thought and the brain to philosophy is not obvious.

Gall's ideas (that current brain imaging works are following) is that we should find in the brain "as many special organs as there are inclinations, feelings, faculties that basically differ among one another."

Fig. 7.2 Human skull annotated by Pastor Oberlin according to the Gall system

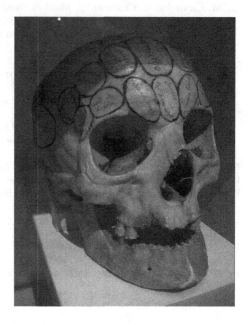

Gall quickly generalized the idea of determining around thirty organs in the brain: the organs of physical love, of friendship, of the metaphysical spirit, and so on. This theory proved to be erroneous, but it was one of the first to assign a localization to the different brain functions (Fig. 7.2). Regardless of the question of accuracy in phrenology, the "brain localization" idea remained controversial for a few years, as some researchers in favor of "brain holism", refused to admit that the brain could be separated into separate functioning units.

"I am told that such a brain area allows me to do such a thing. I do not like the idea of my brain allowing me to do anything, even though I am obviously willing to acknowledge that material conditions are necessary."

According to the author, it is as if there were a "fight" between thought and the brain.

Emmanuel Fournier set himself two objectives:

1° to see how techniques can make the brain speak, and how the brain can actually end up saying what it is made to say with these techniques.

2° to see thoughts in the brain as we ended up seeing diseases in the body, and no longer see the conclusion like wild, free essences.

References

1. L. Alexandre, *La Guerre Des Intelligences: Intelligence Artificielle Versus Intelligence Humaine* (J.C, Lattés, 2018)
2. E. Charles, Bouée and François Roche. *La chute de l'Empire humain: Mémoires d'un robot* (Grasset, 2017)
3. M. Crevoisier, C. Chaineau, C. Malabou, Note sur la stigmergie. *Annales littéraires de l'Université de Franche-Comté*, pp. 140–156 (2018)
4. E. Dutton, D. Van der Linden, R. Lynn, The negative flynn effect: a systematic literature review. Intelligence **59**, 163–169 (2016)
5. I. Filliozat, *L'intelligence du coeur* (Poche, 2013)
6. J.R. Flynn, *What is Intelligence?: Beyond the Flynn Effect* (Cambridge University Press, Cambridge, UK ; New York, 2007)
7. E. Fournier, *Insouciances du cerveau précédé de Lettre aux écervelés*. éditions de l'éclat (Philosophie imaginaire, 2018)
8. H. Gardner, *Les Formes de l'intelligence* (Odile Jacob, 2010)
9. C. Goldin, L. Katz, *The Race Between Education and Technology* (Harvard University Press, 2010)
10. F. Jacob, *La logique du vivant, une histoire de l'hérédité* (Gallimard, 1970)
11. M. Klopfenstein, *Les dimensions de l'intelligence artificielle* (2018)
12. C. Lexcellent, *Comportements linéaires ou non linéaires des matériaux solides: cours et exercices* (Cépadues, 2016)
13. C. Malabou, *Métamorphoses de l'intelligence. Que faire de leur cerveau bleu?* (PUF, Paris, 2017)
14. Edgard, Morin and Massimo Piattelli-Palmarini. *Le Cerveau humain* (Paris Seuil, 1978)
15. K. Robinson, *L'élément* (PlayBac, 2009)
16. G. Vignaux, La ruse, intelligence pratique. *Sciences Humaines* (2003)

Chapter 8
The Truth of "Artificial Intelligence": The Disappearance of the Body

Abstract Intelligence does not reside solely in the brain, as neuroscience would sometimes have us believe! Thus, Joël Chevrier, a professor at Grenoble Alpes University doubts that artificial intelligence, even if it is equipped with robots, could be competitive as regards bodily activities (sport, dancing, music, cooking, etc.).

Intelligence does not reside solely in the brain, as neuroscience would sometimes have us believe! Everything in our being participates in our intelligence. I was going to say all our organs! Robots do not have bodies as such.

Thus, Joël Chevrier, a professor at Grenoble Alpes University [1] doubts that artificial intelligence, even if it is equipped with robots, could be competitive as regards bodily activities (sport, dancing, music, cooking, DIY, etc.).

He mentions the movie "Her" by Spike Jonze. In this film, artist Scarlett Johansson, aka Samantha, is only present through her voice, she is a chatbot, a robot who has learned to hold a conversation. She has no body, she is artificial intelligence. She is interacting with a male character, who is a human being. They can exchange constantly, and at first glance one might think that it "works". But on closer inspection, we realize that AI only interacts with a limited part of what makes up a human being, which could be called the language capacity. And this does not bother the male character, who does not do much with his body; he loves his home and does not practice any physical or manual activity. What interaction would be possible if AI were asked to "exchange" in other fields than language, that can be digitized?

Thus, AI is ultra-efficient in non-embodded intelligence, the intelligence that memorizes and compares large amounts of data. Thus, many professions are better performed, or will be better done by bots, robots, for example, to establish a diagnosis, or choose the most suitable treatment for a patient. And just as the robot wins at the game of Go or at chess, it is better ranked in a competitive entrance examination whose tests are based on the ability to accumulate a great amount of information.

But not all information is digitizable, and Joël Chevrier wishes to show that intelligence affects many other fields.

And to illustrate this, he asks the following question: can an AI be a mountain dweller, just like a student from Grenoble, for example? Can it be a climber, a skier,

C. Lexcellent, *Artificial Intelligence versus Human Intelligence*, SpringerBriefs in Applied Sciences and Technology, https://doi.org/10.1007/978-3-030-21445-6_8

a good physics student and possibly a musician all at the same time, just the same as some students?

"In the face of this human versatility, AI can wait for some time", he says.

Thus, the human being is also above all a body, a being that acts and creates using his/her body, composes with the real, a body in movement, a body that defines "our way of being in the world". And as long as we keep on exploring all these facets, we are not exchangeable with software. But we will have to be wary because the school more often targets the heads than the bodies.

To conclude this chapter, we can resume the words of Michel Serres when he described his book "Variations on the body" [2]:

"Written in praise of physical education teachers and coaches, mountain guides, athletes, female dancers, mimes, clowns, craftsmen and artists ..., these Variations describe the admirable metamorphoses that their bodies can accomplish. Animals lack such a variety of gestures, postures, and movements; flexible to the point of fluidity, the human body imitates things and living things at leisure; moreover, it creates signs. Already present in these positions and metamorphoses, the spirit then emerges from these variations. The five senses are not the only source of knowledge: knowledge emerges to a large extent from the imitations made possible by the plasticity of the body. In the body, with it, and through it, knowledge begins. From sport to knowledge, it goes from forms to signs, to take off in the form of a glorious body. What is incarnation? A transfiguration."

References

1. J. Chevrier, L'intelligence est dans le "faire". The Conversation (2017)
2. M. Serres, *Variations sur le corps* (Le Pommier, 2013)

Chapter 9
We Will Soon Be Able to Hack the Human

Abstract In a decade or two, the historian says, a world in which we can be hacked and manipulated like a machine will come. How shall we live if a regime or a borderless company finally knows us better than ourselves and precisely identifies our desires, our fears, our weaknesses? Submission to algorithms sketches an alarming world in which the notions of freedom and free will be radically challenged.

First of all, no one can build a wall against artificial intelligence or climate change. International cooperation will be essential if we want to succeed in regulating artificial intelligence and biotechnology. For example, if China or Russia acquire armed robots, it will be the beginning of an arms race. Five years ago, no one except perhaps China understood the potential of artificial intelligence, which might lead to the disappearance of humanity [2].

Liberal democracy assumes that our brain is a black box filled with our desires and thoughts, to which we alone have access. Yet, the current crisis arises from the fact that "technologies for hacking people" are available, not just laptops and computers. The revolution lies in the fact that your desires, your feelings, your thoughts can be understood so as to control you.

How can we get out of that trap?

AI and big data are certainly used by companies to better understand people and manipulate them. But the same technologies can be used to help people better understand one another and develop their immunity. We are vulnerable today because we are not aware of our weakness. That is why we must give up the illusion of total free will: what we think and feel depends on the internal processes of our body and our brain, which we know nothing about and do not control.

In the twentieth century, we had the following mantra: "trust yourself and follow your heart." But your heart may be a Russian agent! This does not mean that no one can be trusted. But we must understand that we are vulnerable to manipulation [2].

A natural question is who should control the databases? the state or the companies? It is a choice between two evils!

Another natural question about education is what should be the use of school?

Nobody knows what jobs will be like in 30 years. It is a waste of time and resources to pretend that we are preparing young people to them. The best bet is to teach children

C. Lexcellent, *Artificial Intelligence versus Human Intelligence*, SpringerBriefs in Applied Sciences and Technology, https://doi.org/10.1007/978-3-030-21445-6_9

how to change. Individuals will have to reinvent themselves several times in their life. Most routine jobs, not necessarily physical ones, are bound to disappear. The nurse's job, which requires a lot of manual skills, seems less threatened than the doctor's, who analyzes the data, compares with the antecedents, looks for a pattern. That is exactly what AI will do, and will do it much better.

New jobs will change constantly. AI will bring about a cascade of more and more powerful revolutions. This means that the number-one problem will be reinventing oneself—reinventing not only one's professional identity, but also one's deep identity.

These words by Harari [2] seem to me more constructive than the ones by Alexandre [1] on the same subject.

The philosophy of traditional education is based on building a stable identity, with skills that will be useful throughout life. Quite the contrary, we must think that we will no longer leave the education system.

"We used to build stone houses. Tomorrow, these personalities will have to be tents" [2].

For this restructuring of education to be carried out, the effort must come from the State. But psychological barriers may be higher than economic ones. Many people will be unable to reinvent themselves. We must therefore accept the fact that more and more people will be there who will live decades after the loss of their jobs, and that they will have to be helped financially.

What a sacrilege! Mindsets will have to change about the universal income.

The biggest problem will be for countries like Bangladesh, Nigeria or Vietnam, whose economy is based on a cheap labor force, because the technological revolution will allow factories to return to California or France. Using 3D printers, T-shirts will be manufactured for much less in New York than in Bangladesh. Computer codes will replace textile workers.

Yuval Noah Harari is an Israeli, so he talks about his own country. "Let us consider AI: one of the most important laboratories of the formation of a digital dictatorship is found in the occupied West Bank. How can a population of 2.5 million people effectively be controlled using AI, big data, drones, and cameras? Israel is the leader in the field of surveillance: the country carries out experiments, and then exports them all over the world."

In a decade or two, the historian says, a world in which we can be hacked and manipulated like a machine will come. How shall we live if a regime or a borderless company finally knows us better than ourselves and precisely identifies our desires, our fears, our weaknesses? Submission to algorithms sketches an alarming world in which the notions of freedom and free will be radically challenged.

References

1. L. Alexandre, *La Guerre Des Intelligences: Intelligence Artificielle Versus Intelligence Humaine* (J.C, Lattés, 2018)
2. N. Harari, Yuval, *21 lessons pour le XXIe siècle* (Albin Michel, 2018)

Chapter 10
Conclusions

Abstract In spite of multiple definitions, the term "intelligence" is to be taken, very cautiously. There is a misunderstanding between strong Artificial Intelligence (Singularity...) and Human Intelligence advocated by the philosophers.

First of all, owing to the multiple definitions we provided, the term "intelligence" is to be taken very cautiously.

In short, it is connoted. Someone qualified as "stupid" (or non-intelligent) can be considered so especially if he/she is not an intellectual (sic), but can have a valuable form of manual or practical intelligence.

Then artificial intelligence comes up in the media, as if bodiless intelligence could exist [2], without a consciousness, without an unconscious, without thoughts [4].

As long as we contemplate weak artificial intelligence in which the machine does not make a decision instead of a human, it may be appropriate, and of course beneficial. But where will the worker or the handler fit in?

In France, the number of jobs lost owing to robotization is estimated to be 2 million, but as Alexandre [1] would say, a robot works round the clock, is not a member of a trade union, does not go on strike, unlike a human being who has a limited working time (around 35 h a week). So it would be necessary to rethink the labor market and the required skills.

AI can control the behavior of any individual in China (running a red light, for example).

In old people's homes, in caring for the elderly, I am not sure that robots can replace human contact. To truly speak involves listening to what the other person says, exchanging, in short, having human relationships.

Strong artificial intelligence apparently poses more problems, as if the machine had run out of human control.

This paves the way for all kinds of fantasies, to implanting silicon chips in the brains of a caste of privileged ones.

Cognitive neuroscience researcher Stanislas Dehaene recently presented his new book "Learn!" published by Odile Jacob [3]. Can school programs be inspired from

C. Lexcellent, *Artificial Intelligence versus Human Intelligence*, SpringerBriefs in Applied Sciences and Technology, https://doi.org/10.1007/978-3-030-21445-6_10

neurosciences? How can we allow children to regain concentration capacities? What knowledge is essential for a teacher to design an effective educational program?

He thinks that a child's algorithm is superior to any algorithm generated by artificial intelligence. However, Ricoeur [7] wrote a rather sharp criticism of neurosciences which appeared too "technicist" to him, not human enough. According to neurosciences, all of our mental states can be tracked down to the activity of our neurons, that is, to chemical reactions. According to Suzanne [8], neurosciences occupy a reductionist position as regards the way they are defended by neurobiology, which consists of putting consciousness and neurons on a par.

The great danger today is to be deluded by strong artificial intelligence and its avatars and the takeover by GAFA.

Let us recall the words of Ray Kurzweil (the chief engineer of Google) "From the 2030s, we will, thanks to the hybridization of our brains with nano-electronic components, have a demiurgic power."

"Singularity is close," Ray Kurzweil wrote in 2005 [5]. Singularity is that moment when the intelligence of machines will overcome human intelligence.

Do not panic ! Human intelligence is the fruit of the brain and in particular of thought, of imagination. Finally, does intelligence in the absence of a body exist?

Since robots do not have a body, they will never have "common sense". They merely cannot be programmed for that.

It is necessary to think the human being at the time of the "increased man" [6]. Let us not forget that the machine is based on calculations, it can simulate emotions, but cannot feel them. Let us remember that an emotion is a form of vulnerability intrinsic to human beings.

In summary, AI poses very concrete ethical problems, and answers are still in their infancy.

Two interesting applications of AI have been developed in our FEMTO-ST laboratory, namely, reservoir computing and AI for predicting giant waves.

And what if finally artificial intelligence and human intelligence were in no way related because they do not evolve in the same field!

References

1. L. Alexandre, *La Guerre Des Intelligences: Intelligence Artificielle Versus Intelligence Humaine* (J.C, Lattés, 2018)
2. J. Chevrier, L'intelligence est dans le "faire". The Conversation (2017)
3. S. Dehaene, *Apprendre! Le talent des cerveaux, le défi des machines* (Odile Jacob, 2018)
4. E. Fournier, *Insouciances du cerveau précédé de Lettre aux écervelés*. éditions de l'éclat (Philosophie imaginaire, 2018)
5. R. Kurzweill, *The Singularity is Near: When Humans Transcend Biology* (Penguin New York, 2005)
6. T. Magnin, *Penser l'humain au temps de l'homme augmenté face aux défis du transhumanisme* (Albin Michel, 2017)
7. P. Ricoeur, *La mémoire, l'histoire, l'oubli* (POINTS, 2000)
8. E. Suzanne, La psychanalyse face aux neurosciences. Implications philosophiques (2009)

Printed in the United States
By Bookmasters